Climate change, environmental impact and the limited natural resources urge scientific research and novel technical solutions. The monograph series Green Energy and Technology serves as a publishing platform for scientific and technological approaches to "green"—i.e. environmentally friendly and sustainable—technologies. While a focus lies on energy and power supply, it also covers "green" solutions in industrial engineering and engineering design. Green Energy and Technology addresses researchers, advanced students, technical consultants as well as decision makers in industries and politics. Hence, the level of presentation spans from instructional to highly technical.

Indexed in Scopus.

Indexed in Ei Compendex.

Kaliappan Sudalyandi · Rajeshbanu Jeyakumar

Biofuel Production Using Anaerobic Digestion

Springer

Kaliappan Sudalyandi
Department of Civil Engineering
College of Engineering
Guindy, Anna University
Chennai, Tamil Nadu, India

Illinois Institute of Technology
Chicago, USA

Rajeshbanu Jeyakumar
Department of Biotechnology
Central University of Tamil Nadu
Thiruvarur, Tamil Nadu, India

ISSN 1865-3529   ISSN 1865-3537 (electronic)
Green Energy and Technology
ISBN 978-981-19-3745-3   ISBN 978-981-19-3743-9 (eBook)
https://doi.org/10.1007/978-981-19-3743-9

This Springer imprint is published by the registered company Springer Nature Singapore Pte Ltd.
The registered company address is: 152 Beach Road, #21-01/04 Gateway East, Singapore 189721, Singapore

Preface

It is waste and waste everywhere, discharged from processes that were mandated to draw energy and elevate the comfort zone and luxury-life style of human beings, in this twenty-first century. These wastes, in the form of gas, liquid, and solid, contaminate enormously the atmosphere, land, and water bodies.

The energy-driven attempt of the modern world exploits the mother earth, exhausts the natural energy resources, and pushes the world population into severe health risks, as could be seen from present disasters in the form of global warming, climate change, and the spread of ill-health among the world population. Should we all simply be spectators of these disastrous evolutions or do something that could pave some solutions?

The authors of this book are working in the field of environmental engineering, attempting towards clean energy generation, eco-friendly waste disposal, and guiding the younger generation to work on these issues to provide better solutions for the past 30 years with international experience. The thirty years plus experience encouraged me to write this book towards answering the questions raised above.

This book on biofuel production from biomass as the core substrate is aimed at providing solutions to generate clean/green energy. Generating biofuels from renewable feedstock will be the best way to balance the current world energy demand and strengthen the global economic crisis. These biofuels are green energy resources and a primary alternative to fossil fuels. In transportation, biofuels are blended with existing fuels such as diesel and gas. Blending biofuel with others is an effective way of reducing carbon emissions during transportation.

This book elucidates essential concepts, mechanisms, applications, and useful outcomes of biofuel generation processes. To understand the concept easily, over 50 illustrations (graphs, diagrams, and photographs) and 25 tables with data relevant to most technical parts of the biofuel production such as reactor type, cost, and net energy balance are provided. The other attractive feature of this book is that it covers cost and energy analysis of the biofuel production process.

This book can serve as a resource material for professionals, researchers, and students, as well as readers who do not have much knowledge of the waste to biofuel (WTBF) process and its associated applications.

Chennai, India   Kaliappan Sudalyandi, Ph.D.
                                                                  (Illinois, Tech.)
Thiruvarur, India   Rajeshbanu Jeyakumar, Ph.D.

Contents

About the Authors

Dr. Kaliappan Sudalyandi is a Professor of Emeritus, Department of Civil Engineering at Anna University, Chennai. He has previously served as the Vice-Chancellor of Anna University Tirunelveli, and Director of the Institute of Remote Sensing, College of Engineering Guindy at Anna University, Chennai, Advisor and Dean Ponjesly College of Engineering, Nagercoil and Dean at Thiagarajar College of Engineering, Madurai. He has earned his B.E. (Civil Engineering) and M. Tech. (Remote sensing) from College of Engineering, Guindy, Anna University, and his Ph.D. in environmental engineering from the Illinois Institute of Technology, Chicago, USA. He has contributed to 2 books and over 91 research publications in international and national journals.

Dr. Rajeshbanu Jeyakumar is an Associate Professor in the Department of Biotechnology at the Central University of Tamil Nadu, Thiruvarur, India. He is also a Visiting Professor at the Center for Excellence in Environmental Studies, King Abdulaziz University, Saudi Arabia. He has previously served as visiting researcher at East China Normal University, China and post-doctoral fellow at Sungkyunkwan University, South Korea. He is the author of 300 research publications with a cumulative impact factor of 1300, citation of 10000, and H index of 52. He is listed among the top 2% of scientists of the world in the field of environmental biotechnology by Stanford University researchers.

Chapter 1
Introduction

The term fossil fuel was known to be introduced, for the first time, by the German scientist and author Andreas Libavius in his book *Alchemia* in the year 1597. Fossil fuel became the principal energy source since the beginning of industrialization. During the past few decades, scientific and technological innovation has transformed the global civilization very distinct in its style of functioning by inheriting the innovative application in attending the day-to-day responsibilities. Such a lifestyle demands more energy that is being derived from fossil fuels that account for 80% of all energy.

These fossil-based fuels comprise of coal, petroleum, natural gas, tar, oil shales, and gravel. These products meet the energy demand associated with electricity, transportation, and heating. The greenhouse gases (GHG) have a significant influence on the atmospheric earth, which increases the earth's surface temperature to 34 °C, from a normal temperature of around 20 °C, thereby making the planet with an average increase in worldwide temperature of 14 °C [1]. The greenhouse gases comprise of methane, carbon dioxide, nitrous oxide, and ozone. These gases have the potential to discharge and absorb infrared irradiations based on their specific characteristics and composition. On the other hand, the extensive development of industry-based practices such as the combustion of fossil fuels leads to a severe extent of GHG emissions into the atmosphere. This results in many serious impacts on the Earth, intensifying the atmospheric carbon dioxide and causing global warming and climate change.

Further, the energy demand was not at pace with the energy supply due to limited fossil fuel resources. This over-exploitation of energy was due to the population growth, increased industrial activities and the development of metropolitan infrastructure.

The burning of fossil-based fuels emits detrimental greenhouse gases (GHG) such as carbon dioxide, methane, sulphur oxide and nitrous oxide. These gases led to acid rain, global warming and climate changes [2, 3]. Around 3.3×10^{10} tons of anthropogenic carbon dioxide was emitted globally in 2018 [4]. The present level of carbon dioxide (394.5 ppmv) is projected to rise by 500 ppmv around 2050 if emissions are not controlled [2]. Many technologies have been developed for capturing carbon dioxide, which include adsorption techniques using amines [5], carbonates

© Springer Nature Singapore Pte Ltd. 2022
K. Sudalyandi and R. Jeyakumar, *Biofuel Production Using Anaerobic Digestion*,
Green Energy and Technology, https://doi.org/10.1007/978-981-19-3743-9_1

Table 1.1 Various substrates and their yields

Type	Biogas yield per ton fresh matter (m^3)	Electricity produced per ton fresh matter (kWh)
Municipal solid waste	101.5	207.2
Fat	826–1200	1687.4
Maize silage	200/220	409.6
Chicken litter/dung	126	257.3
Food waste (disinfected)	110	224.6
Horse manure	56	114.3
Pig slurry	11–25	23.5
Fruit wastes	74	151.6
Cattle dung	55–68	122.5
Sewage sludge	47	96.0

[6] and ammonia [7]. In addition, the pre-combustion techniques such as chemical looping [8] or capture route of oxy fuel [9] are insufficient to lock up the emission of carbon dioxide [10]. To solve this, over the years, alternate energy resources such as wind, hydroelectric, solar and renewables are being tried out to meet the future energy requirement. However, the presently available energies would not sufficiently reduce the greenhouse gas (GHG) emissions and the dependency on fossil-based fuels. These imbalances urged the need for an alternative renewable and eco-friendly energy for better survival and sustainable development.

The waste biomass mainly utilized was the non-edible feedstock which is a biogenic waste substrate. This could be a potential alternative for fossil fuels to enhance bioenergy generation. These waste biomasses are rich organic sources utilized as a major energy reserve from ancient times. It meets the energy requirement by generating bioethanol, biobutanol, biomethane, biohydrogen, electricity and heat through various techniques such as combustion, gasification, pyrolysis, digestion, fermentation and extraction, shown in Table 1.1.

Sustainable management of unused waste biomass as a substrate for bioenergy generation promotes renewable and environmentally friendly technological development. It is the only carbon–neutral renewable resource that can be transformed into useful solid, liquid and gaseous biofuels by appropriate techniques. This technique has certain technical challenges such as low energy yield, investment and pollution.

One such technique is Anaerobic Digestion (AD) which is a matured biotechnological process that has an intensive benefit in terms of economic, environmental, and energetic aspects within the scientific communities. Anaerobic Digestion (AD) showed significant progress in the field of research and its application. AD is the microbial degradation of waste biomass both in natural and engineered infrastructure.

The AD process has four stages namely: hydrolysis, acidogenesis, acetogenesis and methanogenesis, which are detailed in Fig. 1.1.

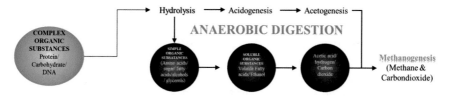

Fig. 1.1 Complex microbial process of AD

AD was carried out by acid-forming bacteria and methanogens through biological reactions. Acid-forming bacteria convert the organic biomass into fragments, and finally to organic acids in an acidic environment. The methanogens promote biomass stabilization to form biogas. Also, the factors such as reactor design, feedstock characteristics and operational condition influence the process efficiency. This AD process can be carried in via both batch and continuous mode. A steady state is never achieved in a batch process, whereas a continuous process produces biogas at a constant rate. Biogenic or Biomass waste is a prime target substance to be used in AD since it has immense prospects in a commercial scale [12].

Consequently, wide usage of biofuels can improve the air quality and is capable of reducing greenhouse gas emissions. The main advantages of biofuels over other conventional biofuels are their sustainability and renewability. Biofuels, in contrast to fossil fuels, release less amount of carbon dioxide gas into the environment. One of the advantages of biofuel is that it can be generated from a variety of waste biomass.

1.1 What is Biofuel?

The term biofuel refers to solid, liquid and gaseous fuels mainly produced from biomass. Biofuels are significant as they hint extensive range of topics such as ecological concerns, energy security and socio-economic problems associated with the rural sector. There are different types of biofuels available, namely, biomethane, biohydrogen, bioethanol, biomethanol, biobutanol, biodiesel, biosyngas, bio-oil, etc.

1.2 Classification of Biofuels

Biofuels are classified into primary and secondary fuels. The biofuels that are utilized in unprocessed form (i.e., for cooking, heating and occasionally for electricity generation) are known as primary biofuels. In primary biofuel production, organic substances are burnt to generate energy. Examples of primary biofuels biomass are wood, wood chips, pellets, fuel wood, unprocessed plant material, etc. The primary biofuel is energy efficient and it resulted in sufficient energy production per kilogram of biomass than liquid-based fuels. However, primary biofuels are not environmentally safe as it troubles the ecological niche by the emission of toxic substances, gases, etc. For instance, the burning of solid fuels such as wood and unprocessed plant materials leads to air pollution, as the fuels are not refined. This is mainly due to the partial combustion of wood. Besides, these fuels are inappropriate for transportation of light vehicles as burning is cumbersome. The biofuels, which are generated by processing the biomass are called secondary biofuels. These biofuels can be utilized in vehicle transportation and industrial processing. Based on the raw materials (biomass—crops, plants, agro residues, food waste and domestic wastes, etc.,) utilized and the technology used for biofuel production, the secondary biofuels are further classified into four types, namely, first- generation, second-generation, third-generation and fourth-generation biofuels.

1.3   First-Generation Biofuels (FGB)

First-generation biofuels (FGB) or conventional biofuels are primarily produced from edible food substances such as sugar (sugarcane), starch (corn), wheat grains, (oil seeds) vegetable oils and purified animal fats. These biofuels are generated by the action of microbes and their enzymes via the fermentation of starch, sugar and vegetable oils. Examples of FGB are biogas, biodiesel and bioalcohols (butanol, propanol, methanol). As FGB utilizes edible food items as feedstocks, these biofuels are banned in European Union. Among the FGB, bioethanol and biodiesel are the two major types of biofuels that are generated commercially. Biodiesel is produced via transesterification of vegetable oils and residual oil as well as fat. These biofuels can be used as an alternative to diesel with slight modifications in engines. Ethanol also called as ethyl alcohol, alcohol or grain alcohol is a volatile fuel, flammable and colourless liquid. It is an alternative to petrol (gasoline). It is mainly produced from starch and sugar (extracted from cane sugar, beet molasses, miscanthus, grain, switch grass, sweet potatoes, etc.) through fermentation. The sequential steps involved in bioethanol production are the fermentation of sugars by microbes, distillation, dehydration and denaturing (if essential). Before fermentation, few crops need prehydrolysis and saccharification process to hydrolyze starch into sugars and are mediated by enzymes. First-generation biofuels are energy efficient and upon burning, emits less amount of toxic gases. The disadvantages of these biofuels are associated with their competition for land and water with cultivational crops. Increased production and processing cost often need government grants to strive with petrol-based products.

1.4 Second-Generation Biofuels (SGB)

Second-generation biofuels (SGB) are also called as "olive green" or "cellulosic-ethanol" fuel. The primary feedstocks that are utilized for the production of second-generation biofuels are non-edible food crops such as lignocellulosic crops, waste vegetable oils, agricultural residues (such as rice straw, wheat straw, paddy straw, etc.) and industrial residues. Second-generation biofuels are mainly derived from lignocellulosic crops. The production process of second-generation biofuels supplies a sufficient quantity of sustainable biofuel with increased ecological benefits when compared to first-generation biofuels. Examples of second-generation biofuels are biohydrogen, bioethanol and biomethane. The biohydrogen and bioethanol generation capability from lignocellulosic crops relies on disintegration methods used, microbial culture or enzymes and type of biomass. Pretreatment of lignocellulosic biomass improves the porosity, remove inhibitors and biodegradability. Due to pretreatment, the accessibility of cellulose is increased and get converted into simpler forms of sugars via enzyme-mediated hydrolysis during fermentation. The pretreatment enhances biohydrogen production from lignocellulosic crops.

1.5 Third-Generation Biofuels (TGB)

Third-generation biofuels are also called as "algal fuel" or oilage" as they are generated from algal biomass. Examples of third-generation biofuels are biodiesel, biobutanol, biopropanaol and bioethanol. Cultivation of TGB (algal) biomass aids in sustaining ecological balances by utilizing the carbon dioxide that exists in the atmosphere. Algal biomass (both macroalgae and microalgae) possess a considerable amount of lipids when compared to feedstocks utilized for conventional biofuels.

1.6 Fourth-Generation Biofuels (FGB)

Fourth-generation biofuels are the biofuels that are generated from plant biomass in which the biomasses are metabolically engineered to enhance carbon capture and energy storage. In this biofuel generation process, the plants, trees and algal biomass are engineered as carbon capture apparatus and enhance the lockage of carbon in various parts of biomass such as stem, leaves, branches, etc. This, in turn, resulted in enhanced production of biofuel.

1.7 Global Status of Biofuel Production

The greater utilization of biofuel turns many countries to reduce the usage of petroleum-based fuels. As a result, the rural economy and quality of air are improved. Figure 1.2 represents the Global scenario of biofuel production [11].

Among the countries, United States and Brazil are leading biofuel producers, particularly the biofuel bioethanol. Besides, countries like China, India, Canada, Thailand, Argentina and Columbia have also established biofuel production. International Energy Outlook 2020 has stated that biomass is the rapidly growing energy source for biofuel production at the global level. The utilization of biofuel production may increase by 2.8% annually and is expected to increase around 15% by 2035 in the global energy market. The report also highlights the importance of switching forward to liquid biofuels. The liquid biofuels may fulfil 29% of global energy requirements, nearly 5 million barrels/day. It has been expected that 75% of the advanced biofuels (blend of bioethanol and other advanced bioalcohols) will be generated from celluloses (second-generation and third-generation biomass). In future, biofuel production will be responsible for sustainable energy development in technologically advanced and developing nations. For instance, in the United States, biofuels are the major supplier to the industry sector and are anticipated to increase from 0.2 to 0.9 quadrillion Btu by 2030. In Brazil, the usage of biofuel has exponentially increased and it became the world leader in biofuel generation via government directives and policies. In addition, most of the petrol sold in Brazil consists of 20 to 25% of ethanol blend. The rapid requirement for biofuel is also initiated in the US where they set their generation target to 21 billion gallons of cellulose-based bioethanol and other advanced biofuels such as biobutanol in 2023. Biofuels generated from first-generation food crops are not commercially feasible

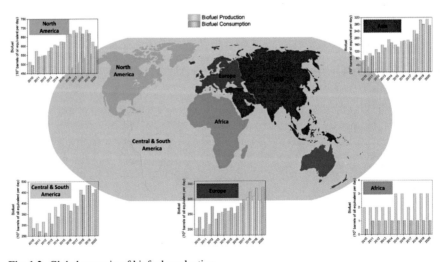

Fig. 1.2 Global scenario of biofuel production

because an increment in fuel price would minimize agricultural productivity. For instance, a 25% increase in fuel price would lead to a reduction in the global food supply. If such a situation occurs, then countries like China, India, Southeast Asia and Europe would face a heavy reduction in food supply by about 3.5–6.1%. Hence, substantial effort has been made to generate biofuels from second-generation biofuel resources such as wheat straw, corn stover and other agro residues.

1.8 Significance of Biofuels

- Biofuel being a suitable alternative to fossil and nuclear-based fuels has a low maturation period and less transportation cost.
- It is sustainable and will never expire
- Biofuel facilities need less care when compared to conventional generators
- Biofuel being generated from biomass, considerably minimizes operational cost
- Biofuel projects may fetch economic profits to rural areas.
- The emission of GHG and other toxic substances is reduced.
- The biofuel generation process is cost-effective and less labour intensive.

1.9 Overview of this Book

Biofuel production is considered as an eco-friendly alternative to conventional fossil-based fuel and it meets the future energy demand at the global level. Biofuel production from biomass sources affords significant energy efficiency and profitability over existing conventional fuel production technologies. Though this text focussess mainly on biofuel production from anaerobic digestion perspective, other processes such as Photo fermentation.

This book summarizes the basics, principles, processes, and mechanisms of biofuel production technologies from various biomass. In this chapter introduces the climate changes and impacts of fossil fuels, significance of biofuels, classification of biofuels on the basis of biomass and global scenario of biofuel production. Chapter 2 reviews the hydrolysis process and hydrolysis of various biomass. Chapter 3 covers the strategies to enhance hydrolysis, basics of phase-separated pretreatment, inhibitor removal and significance of liquefaction Chap. 4 outlines the methods used for assessment of hydrolysis particularly based on liquefaction, energy and cost. Chapter 5 covers the various kinetics and modelling of hydrolysis process and its significance. Chapter 6 outlines the basics and theory of fermentation, factors affecting fermentation, strategies employed to enhance fermentation, kinetics and modelling aspects of fermentation, Chapter 7 covers the biofuel generation- methane, hydrogen, biodiesel and bioethanol process, factors influencing biofuel production, and the strategies employed to enhance biofuel production. Chapter 8 explains in detail about types of digestion process for biofuel production (single-stage and two-stage), reactor configurations, microbial pathways involved, process parameters, economics and scale-up, industrial production of biofuel, current problems in industries and solutions.

This book covers all aspects of biofuel production methodologies and is aimed to attract all the readers mainly undergraduates, postgraduates, research scholars and engineers to understand the basic concepts, theory, methodology and modelling and economics of the biofuel production process. We hope this book will be interesting and informative to all the readers, particularly for their research pursuits.

References

1. Wolfe ML, Ting KC, Scott N, Sharpley A, Jones JW, Verma L (2016) Engineering solutions for food-energy-water systems: it is more than engineering. J Environ Stud Sci 6:172–182
2. Li Y, Park SY, Zhu J (2011) Solid-state anaerobic digestion for methane production from organic waste. Renew Sustain Energy Rev 15:821–826
3. Mata-Alvarez J, Dosta J, Romero-Güiza MS, Fonoll X, Peces M, Astals S (2014) A critical review on anaerobic co-digestion achievements between 2010 and 2013. Renew Sustain Energy Rev 36:412–427
4. Yang L, Ge X, Wan C, Yu F, Li Y (2014) Progress and perspectives in converting biogas to transportation fuels. Renew Sustain Energy Rev 40:1133–1152
5. Sheets JP, Yang L, Ge X, Wang Z, Li Y (2015) Beyond land application: Emerging technologies for the treatment and reuse of anaerobically digested agricultural and food waste. Waste Manag 44:94–115
6. Yi J, Dong B, Jin J, Dai X (2014) Effect of Increasing Total Solids Contents on Anaerobic Digestion of Food Waste under Mesophilic Conditions: Performance and Microbial Characteristics Analysis. PLoS ONE 9:e102548
7. Guo X, Wang C, Sun F, Zhu W, Wu W (2014) A comparison of microbial characteristics between the thermophilic and mesophilic anaerobic digesters exposed to elevated food waste loadings. Bioresour Technol 152:420–428
8. Stolze Y, Zakrzewski M, Maus I, Eikmeyer F, Jaenicke S, Rottmann N, Siebner C, Pühler A, Schlüter A (2015) Comparative metagenomics of biogas-producing microbial communities from production-scale biogas plants operating under wet or dry fermentation conditions. Biotechnol Biofuels 8
9. Sasaki D, Hori T, Haruta S, Ueno Y, Ishii M, Igarashi Y (2011) Methanogenic pathway and community structure in a thermophilic anaerobic digestion process of organic solid waste. J Biosci Bioeng 111:41–46
10. Lin L, Yu Z, Li Y (2017) Sequential batch thermophilic solid-state anaerobic digestion of lignocellulosic biomass via recirculating digestate as inoculum—Part II: Microbial diversity and succession. Bioresour Technol 241:1027–1035
11. Outlook for biogas and biomethane , World energy outlook special report 2000 , International Energy Agency
12. Lin L, Yu Z, Li Y (2017) Sequential batch thermophilic solid-state anaerobic digestion of lignocellulosic biomass via recirculating digestate as inoculum – Part II: Microbial diversity and succession. Biores Technol 241:1027–1035

Chapter 2
Biomass for Biofuel Generation

2.1 Overview

Biomass originates from organic material and can be used to generate energy. It is considered to be a renewable and sustainable source of energy because of the availability of biomass residue throughout the year, such as scrap wood, waste activated sludge and municipal solid waste. Biomass waste is the only resource of biofuel for domestic use in most developing nations. The approximate biomass production in the world is 104.9 petagrams (104.9×10^{15} g—about 105 billion metric tons) of carbon/year. In developing countries, agriculture plays a major role in the development of its economy. Every year, a larger amount of residues are obtained from the crop during its harvesting. Rice, wheat, sugarcane, corn and groundnut are examples, which generate a significant quantity of residues. For industrial purposes and domestic uses, these residues are considered as raw materials for energy production. This chapter discusses biomass classification, types of waste and physico-chemical characteristics of various biomass feedstock for waste to energy (WTE) conversion.

2.1.1 Classification of Waste Biomass

The waste biomass is mainly classified into high, medium and low potential waste based on its calorific value. Calorific value is an important factor in estimating the quantum of energy produced from waste biomass through a combustion process. Utilization of biomass as a fuel depends upon the amount of burnable organic matter present in it. The calorific value of a substance depends upon the amount of carbon present in it.

© Springer Nature Singapore Pte Ltd. 2022
K. Sudalyandi and R. Jeyakumar, *Biofuel Production Using Anaerobic Digestion*,
Green Energy and Technology, https://doi.org/10.1007/978-981-19-3743-9_2

2.1.2 High Potential Waste Biomass

These are defined as the waste biomass that has high calorific values ranging from 24 to 40 MJ/kg [1]. Examples of high potential waste biomass include industrial waste, algal biomass and biomedical waste.

2.1.3 Medium Potential Waste Biomass

These are defined as the waste biomass that has medium calorific values ranging from 19 to 22 MJ/kg [1]. Examples of medium potential waste biomass include agricultural residues, food wastes, municipal solid waste (MSW) and forest residues.

2.1.4 Low Potential Waste Biomass

These are defined as the waste biomass that has low calorific values ranging from 2 to 4.3 MJ/kg [2]. Examples of low potential waste biomass include animal wastes—cattle manure, pig and poultry manure, etc.

Based on the physical property, the wastes are classified into two types, namely, liquid and solid wastes.

2.1.5 Liquid Wastes

Various wastes that exist in solid form can also be transformed into a liquid form for clearance. It comprises direct and indirect origins, for instance, squall and wastewater. These comprise swab water from houses, wastewater generated during the processing of raw materials in the industry and liquid used for washing in production and manufacturing units.

2.1.6 Solid Wastes

Solid wastes primarily include trash, refuse or debris that originates from our residences and erstwhile areas. These comprise groceries waste, activated sludge cake, food waste, etc. It can include waste that exist in non-liquid form.

2.2 Types of Biomass Wastes and Their Constituents

Biomass waste originates from various sources such as agriculture, livestock farming, forestry, industrial, urban, hospital, etc. Based on the origin of biomass waste, it is further classified into eight types. In this section, a comprehensive discussion about various biomass types and their constituents has been detailed.

2.2.1 Agricultural Residues

Agricultural residues are the materials left during the harvesting and collection of agricultural crops. These residues include husks, seeds, bagasse, molasses and roots. During corn harvesting, the leaves and stems that are not utilized are the biomass residues. Due to their high carbohydrate content, these residues are regarded as a suitable raw material to generate biofuels. Globally, the amount of agricultural residues utilized for biofuel generation was estimated to be 249 million tonnes/year. This amount of residues can approximately produce 355 kilotonnes of biofuel.

2.2.1.1 Rice Straw

In earlier days, rice straw has been mainly utilized for cattle and animal feeds, as a packing material and for heating and cooking purposes. In many countries, rice straw is utilized as an organic fertilizer by charring the straw in the field with ash [3]. Comparatively, minor amounts are utilized as a raw material in paper, board and building industries, and for rearing farm animals as its fodder and bedding. Recent techniques such as anaerobic digestion, combustion, gasification and pyrolysis are considered as contemporary technologies and are used to generate heat and electricity [4], bioenergy [5], from rice straw. In developing countries, biomass energy is mainly utilized for cooking and electricity purposes [6].

2.2.1.2 Rice Husk

Husks are considered as loose biomass. In large rice mills, in order to get rid of these rice husks, it is habitually charred so that the power required for rice mill operation is equalized. About 30 to 50% of husks are considered to be utilized by the rice mills. The remaining 50 to 70% can serve as a raw material for the generation of energy. Numerous alternatives have been made by researchers to convert rice husk into value-added products like briquettes and pellets [7]. Ethanol production from rice husks is also an upcoming trend and many researchers have been working on it to attain higher yield [8, 9].

2.2.1.3 Wheat

Wheat plants, which are composed of stem, stalks, straw and dry leaves are significant feedstock for the generation of bioenergy due to their sustainability, abundance and low cost [10]. Fractionation of ethanol from wheat straw is done by pretreating wheat straw by various methods such as enzymatic treatment [11] and ultrasonication [12]. Co-digestion is the preferable method for achieving higher biogas yield in the anaerobic digestion of wheat straw [13]. In paper industries, wheat straw was successfully recycled and about 80% of copy paper is made up of wheat straw. Fibre present in the wheat act as a renewable and eco-friendly biomass, which, in turn, helps farmers form a new market for what remains after the grain is harvested. Wheat straws mainly consist of calcium (Ca), magnesium (Mg), sulphur (S) and micronutrients. In Illinois, USA, wheat straw is utilized along with the soil native for the supply of these additional nutrients, so there is no need to supply these nutrients through fertilizers. A cation exchanger prepared from wheat straw is used to decolourize the distillery spent wash [14].

2.2.1.4 Corn

Among the various agricultural waste, corn straw was considered as most important in terms of production. As much as 90% of corn straw is left unused in the field. About 70% of corn straw consists of cellulose or hemicellulose and dense carbohydrates which are interlocked in chains. Corn straw consists of leaves and stalks of maize plants (*Zea mays* sp. *mays* L.), which are left in a field after their harvest [15]. Corn straw was mainly as a feedstock for ethanol production [16]. Producing ethanol from corn grain and corn straw at the same location can reduce energy consumption and cost. A steam explosion method act on the corn waste and discharges the carbon-based sugars and additional compounds. These compounds are utilized by the microbial fuel cells for the generation of electricity [17].

2.2.1.5 Sugarcane

Sugarcane (*Saccharum officinarum*) is one among the few species of genus *Saccharum* which is utilized as biomass. Sugarcane is crushed to obtain the sugar juice and, in turn, fibrous material is left over which is known as bagasse. Press mud and bagasse are the by-products of sugar industries, which are used to improve chemical, physical and biological properties of soil and enhance the crop quality and yield, therefore, cutting down the chemical fertilizer requirement [18]. Multidisciplinary approaches to handle sugar industries wastes include ethanol production, biogas generation, recovery of chemicals and use of bagasse and bagasse fly ash as adsorbents in water treatment and building materials [19, 20]. Cement industry plays a major role in carbon dioxide emission and considers to be a major contributor to global warming. Sugarcane bagasse ash can be used as a partial replacement for fillers in the cement industry [21]. In building industries, the utilization of man-made substances is extensively used as sound absorbers. The natural fibres obtained from sugarcane bagasse are exploited and used as an auditory substance sound absorber. The worthy auditory description is established at 1.2–4.5 kHz with a typical absorption coefficient of 0.65 [22]. Bagasse is subjected to pyrolysis for the production of biochar [23].

2.2.2 Wastewater Sludge

During the biological treatment of wastewater, waste activated sludge is generated in enormous proportions. This sludge comprises harmful elements, including microorganisms, toxic metals and organic pollutants, that cause odours and hygienic issues. Sludge mishandling and disposal has serious environmental consequences as well as a public health risk. Anaerobic digestion has a great deal of promise in treating the sludge since it eliminates pathogens and odours which stabilizes the sludge and most crucially creates sustainable energy in the form of methane and hydrogen [24]. In the sludge, extracellular polymeric substances (EPS) act as a chelating stratum that develops flocs by clustering microbial cells and obstructing microbe accessibility during hydrolysis.

Protein, carbohydrates, nucleic acid and lipid are the major components of EPS. Loosely bound and tightly bound EPS are often the two kinds of EPS. Due to the obvious intricate floc structure (extracellular polymeric compounds) and rigid cell walls in waste activated sludge, sludge requires long retention durations for digestion which was due to the hydrolysis (yield-limiting phase), this leads to insufficient organic solids breakdown and ineffective methane output and hydrogen. To overcome this, a wide range of techniques are developed for pretreating sludge. Thermal, mechanical, biological, chemical, or hybrid technologies, have indeed been employed to expedite the process of hydrolysis [25]. This pretreatment can help to relieve intracellular compounds by perforating the wall of the microbial cell and making it further susceptible to further bacterial activity. Pretreatment adopted to enhance the AD process should satisfy certain criteria, such as (1) advancement in sugar formation or the ability to produce sugars through enzymatic hydrolysis, (2) preventing carbohydrate breakdown or loss and (3) Preventing the development of inhibitory by-products in the fermentation and hydrolysis processes.

2.2.3 Lignocellulosic Biomass

The polymeric components cellulose, hemicellulose and lignin constitute the majority of lignocellulosic materials. These polymers form a hetero-matrix having fluctuating degrees of linkage and relative composition based upon its kind, species and feedstock [26]. This wide availability of cellulose, hemicellulose and lignin are the crucial elements in identifying the appropriate energy generation strategy for each sort of lignocellulosic material [27, 28]. Lignocellulosic fuel sources necessitate stringent pretreatment that can be processed by commercial cellulolytic enzymes or microbes that generate enzymes to promote the release of sugars during fermentation. Nature's most prevalent polymer is lignin. It can be found in agronomic as well as plant leftovers, and it provides a stiff, effective barrier against pathogen invasion and oxidative stress [29–31].

Lignin is an opaque heteropolymer made up of a network having phenyl propane units (p-coumaryl, coniferyl and sinapyl alcohol) that are bound all together to form linkages. It is considered as "glue" that holds the various elements of lignocellulosic biomass intact, rendering it as a water-repellent substance [3]. For its intimate relationship with cellulose microfibrils, lignin was found to be the main inhibitor in the hydrolysis of lignocellulosic biomass by enzymes and microbes. However, some straws and other leftovers have a high lignin content, which can cause poor biodegradability and limited methane and hydrogen production [33–35]. Pretreating lignocellulosic materials strengthen bioavailability, as well as bioenergy production, to reduce hydraulic retention time [36]. The NaOH pretreated wheat straw yielded 87.5% more biogas and 111.6% more CH_4. Thus, proper pretreatment, nutrient addition and co-digestion of agricultural and waste biomass create a sustainable methane output [37].

Aside from ever becoming a physical barrier, lignin has the following adverse effects: (i) Interference with cellulolytic enzymes which can attach to lignin-carbohydrate complexes and causes ineffective binding; (ii) Hydrolytic enzymes adsorb non-specifically to "sticky" lignin.; and (iii) Toxicology of lignin compounds in microorganisms. To improve the digestibility of biomass, diverse feedstocks have varying levels of lignin, which must be confiscated by various pretreatment processes. The lignin is found to liquefy through pretreatment which gets consolidated upon cooling, shifting its characteristics and can be precipitated. The chemical extraction of lignin which is called delignification causes biomass swelling, structural disturbance of lignin and increased internal surface area. This in turn makes cellulose fibres to be accessible for the cellulolytic enzymes. In addition to delignification, in some cases, pretreatment aids changes in structural characteristics of lignin. As a result, upon anaerobic digestion, when compared to control, pretreated biomass produces more end product.

2.2.4 Micro and Macroalgae

Aquatic biomass is now considered as a viable raw material for biofuel production because it is not in direct conflict with food crops and produces significantly more biomass per hectare than land crops. However, significant economic and technological hurdles must be overcome before such biofuels can be successfully commercialized [31, 32]. Based on its size, they are divided into two: microalgae (unicellular) and macroalgae (multicellular) (Milledge et al. 2014). Algae accumulate carbon dioxide (CO_2) in a short time with high production capacity and produce carbohydrates, protein and lipids (Notoya 2010). For over 50 years, it is proposed to generate biofuels through algae. It is estimated that globally 30.1 million tons of macroalgae are produced in 2016, among them 95% are produced by artificial cultivations and the remaining 5% are produced naturally. Macroalgae, also known as seaweeds, are found as significant organisms in the marine ecosystem where they utilize carbon dioxide and store carbon. Many researchers reported that macroalgae are an effective substrate used for biobased fuel production such as biohydrogen, methane, ethanol and biodiesel. Macroalgae is cultivated on-shore and off-shore. The following factors such as climate, temperature and water saline condition, etc. are considered for commercial macroalgae cultivation. Off-shore macroalgal cultivation is cost-effective, whereas on-shore cultivation needs more cost for processing. In recent years, the usage of algae for biofuel production is gaining attention around the world. Algae offer several advantages as it contains pigments, sugars and lipids, and upon valorization, many commercially important products can be produced from algae. Because of this reason, it is the most commonly employed substrate in biorefinery. Among the substrates used for biodiesel production, microalgae are found to be promising as it exhibits a high growth rate and can accumulate a greater amount of lipids. Usage of microalgae for biofuel production also ensures the food security of a nation as it did not compete with other food crops for land and nutrients. Microalgae can be grown in wastewater; thus, it ensures the water security of a nation and because of the reason it is classified under third- and fourth-generation biomass used for biofuel production. Unlike the first- and second-generation biomass, microalgae did not contain recalcitrant compounds like lignin, which happen to decrease the digestibility of the substrate.

According to the thallus colour, macroalgae are classified into green, red and brown. In green algal classifications, over 4500 species are present which include 3050 freshwater algae species (*Chlorophyceae* and *Trebouxiophyceae*) and 1500 seawater algae (*Ulvophyceae, Dasycladophyceae, Bryopsidophyceae* and *Siphoncladophyceae*). In red algal classifications, *Rhodophyceae* is the main class which includes two sub-classes, namely: *Bangiophycidae* and *Florideophycidae*. The appearance of red colour is due to the presence of pigments such as chlorophyll a, phycoerythrin and phycocyanin. Among the 6000 red algal species found, most of them exist in tropical regions. The brown algae constitutes over 2000 species among which *Phaeophyceae* is the main class and its colour is due to the presence of chlorophyll a and c, b-carotene and other xanthophylls.

Microalgae are tiny creatures that fall into three categories: diatoms, green algae and golden algae. Diatoms are single-celled organisms. Golden algae generate lipids and carbohydrates in the same way that diatoms do [30]. There are several potential microalgae species available for bioenergy production, among the green and blue green microalgae species such as Chlorella, Spirulina, Scenedesmus and Nannochloropsis have gained more attention recently due to their biochemical composition and product yield. It has been discussed in detail in the below sub-section.

2.2.4.1 Chlorella

Chlorella is an autotrophic, unicellular, non-flagellated, microscopic and round-shaped organisms. The cell wall of chlorella contains cellulose and hemicellulose and it provides rigidity to algae. The size of the chlorella varies from 0.2 to 10 μm in diameter. Chlorella has high biomass to productivity ratio due to its autotrophic nature. Chlorella contains high concentrations of biopolymers such as sugars, proteins, carbohydrates and lipids. There are several species under the genus and among them, *Chlorella vulgaris* is most widely used for biofuel generation.

2.2.4.2 Spirulina

Spirulina is an autotrophic, multicellular, non-flagellated, cylindrical and filament-shaped organism. It is classified under the class cyanobacteria. The size of the spirulina varies from 2 to 12 μm, rarely it extends to 16 μm. It multiplies rapidly through binary fission. It contains high amount of protein and carbohydrates, and because of this reason, it is highly exploited in the food industry for the production of single-cell proteins. There are several species under the genus and among them, *Spirulina platensis* is most commonly used for the generation of biomethane.

2.2.4.3 Scenedesmus

Scenedesmus is a genus of fresh green microalgae species, colonial and non-motile in shape with an average diameter of 2–10 μm. It is autosporic species and it is found with multinucleate (four or eight) cells inside a parental mother wall. Scenedesmus is aprotein-rich microalgae. It consists of three major biopolymers, such as protein (8–56%), carbohydrate (10–18%) and lipid (12–14%). It has 72 different species, among them Scenedesmus quadricauda is most commonly suggested for bioenergy production.

2.2.4.4 Nannochloropsis

Nannochloropsis is a genus of green microalgae, non-motile sphere-shaped with an average diameter of 2–3 μm. Often it is found in both fresh and marine environment. It contains a high amount of chlorophyll a with a complete absence of chlorophyll b and chlorophyll c. It has six major species such as *Nannochloropsis gaditana*, *Nannochloropsis granulate*, *Nannochloropsis limnetica*, *Nannochloropsis oceanica*, *Nannochloropsis oculata* and *Nannochloropsis salina*. They are capable of synthesizing high-quality bio-pigments such as astaxanthin, zeaxanthin and canthaxanthin, which are used for several industrial applications. In addition, they have the ability to accumulate high concentration of polyunsaturated fatty acids.

2.2.5 Food Waste

Organic wastes from the food industry come in a variety of forms including Food Waste (FW). These wastes are made up of a variety of organic and inorganic parts. FW has high moisture content and biodegradability due to a significant percentage of kitchen rubbish, leftovers from residences, restaurants, cafeterias, factory lunchrooms and marketplaces [38, 39]. The waste is segregated by source or collected from a co-mingled source which is then separated at a materials recovery plant [40]. FW is easily biodegradable and produces a lot of bioenergy. Food waste contains vegetable tubers, seed remnants and starch which is the most abundant source of carbs. It is available in the form of granules which can have a variety of morphological appearances depending on the plant species, each of which contains several million amylopectin molecules and a substantial number of amylose molecules [41]. Pectins contain methylated and acidic sugars, such as galacturonic acid, and some pectins are highly branched and have a complex structure. These polymers can chelate calcium and form gels [42]. To improve the biodegradability of FW, it must be hydrolyzed. Lignin is a huge, complex molecule made up of cross-linked polymers of phenolic monomers [43–45]. Some of the negative effects of lignin include nonspecific adsorption of hydrolytic enzymes to "sticky" lignin and interference with nonproductive binding of cellulolytic enzymes to lignin-carbohydrate complexes. Lignin derivatives are poisonous to microorganisms [46]. Lignin is present in FW, which should be eliminated by pretreatment to improve biomass digestion [47].

2.2.6 Animal Waste

Animal dung is one of the most used feedstocks for AD. Historically, animal waste was the most common substrate for biogas production in the AD process. In many countries, cattle populations are significant [48]. Livestock generates a lot of manure which is a good substrate for AD. Waste is a mixture of faeces and urine with a chemical makeup that varies greatly depending on the chemical features of the animal's feed [49, 50]. Lignocelluloses, polysaccharides, proteins and other substances are abundant in animal excreta. Cellulases, hemicellulases and ligninases are lignocellulose-degrading enzymes that play a critical role in hydrolyzing lignocellulose into sugars and biofuels [51]. The key difficulty is the complexity of biomass structure, which makes lignocellulosic biomass very resistant to anaerobic breakdown and, as a result, low biomethane output. Biomass recalcitrance refers to native lignocellulose's tenacious anti-degradation qualities, which severely inhibit hydrolysis during the anaerobic digestion process which is the initial step, and impede commercial biomethane generation. As a result, hydrolysis of lignocelluloses to fermentable sugars is the most difficult in developing cost-effective biorefinery feedstock [52–55].

2.2.7 Fish Waste

Solid and liquid wastes are mixed together in fish waste. The solid phase is made up of the fish tissues and bones, while the liquid phase is made up of blood-water and stick-water, which are high in proteins and oils [56, 57]. This type of biological waste is variable in nature so the usage of this kind of waste is limited. These wastes are high in protein (up to 60%), fat (up to 20%) and minerals, such as calcium and hydroxyapatite from bones and scales (up to 20%) [58]. The fish waste streams contain a high amount of palmitic acid, oleic acid and monosaturated acids. When fish waste is digested, it produces a lot of ammonia, which prevents substrate digestion. The accumulation of VFAs can be caused by high ammonia-concentrated acetic acid [59]. The existence of a large amount of oil can hinder the process, depending on the reactor type and rate of organic loading. The co-digestion of two distinct substrates is a technical option developed to minimize the problem [60]. Co-digestion is a relatively new procedure in which two different substrates (co-substrates) are mixed in the reactor to increase organic matter content and hence boost biogas and hydrogen generation rates. Raw materials, co-substrate types and pretreatment methods all influence the composition and yield of biogas. The potential for methane generation is higher in substrates with high lipid content and quickly degrades carbohydrates [61]. The employment of pretreatment procedures improves the intensity of substrate degradation and the efficiency of the process.

2.2.8 Biomedical Wastes

Biomedical waste is also called as contagious waste or medical waste. Biomedical wastes are solids wastes produced at the time of diagnostic process, examination, management and exploration. Biomedical waste comprises tissues, plungers, subsist vaccines, lab materials, sample fluids pointed needles and cultural strains. Biomedical waste could be classified on the basis of its hazardous nature. The inappropriate handling of biomedical waste leads to severe ecological troubles.

2.2.8.1 Classification of Biomedical Waste

The most important aspect of the management of biomedical waste is that the hazardous biomedical waste should not be mixed with the non-hazardous general waste. In order to achieve this, different types of biomedical wastes in hospitals should be collected in different coloured bins and containers. They are classified as follows:

- Yellow bin waste
- Red bin waste
- Black bin waste

- Blue bin waste.

2.2.8.2 Yellow Bin Wastes

It includes human, animal anatomical waste, soiled waste, discarded medicines, chemical waste, microbiology, biotechnology and other clinical laboratory waste. These wastes can be subjected to incineration and plasma pyrolysis.

2.2.8.3 Red Bin Wastes

It includes plastic waste, and disposable items like tubes, catheters, blood or urine bags, gloves, etc. These types of wastes can be subjected to microwaving, hydroclaving and autoclaving. These treated wastes in turn can be sent to registered or authorized recyclers for energy recovery.

2.2.8.4 Black Bin Wastes

It includes chemical solid waste, expired medicines and X radiation waste. These wastes are subjected to incineration.

2.2.8.5 Blue Bin Wastes

These wastes include syringes, needles, scalpels, blades, pasteur pipettes and broken glasses or plastics. Similar to red bin wastes, these types of wastes can be subjected to microwaving, hydroclaving and autoclaving. These treated wastes, in turn, can be sent to registered or authorized recyclers for energy recovery.

2.2.8.6 Characteristics of Biomedical Wastes

The characteristics of biomedical waste are categorized into bio, chemo and radioactive wastes. World Health Organization has proposed that eighty-five percent of hospital wastes are really safe, about ten percent are contagious and about 5% are non-contagious but harmful waste materials. In India, the infectious waste would be ranged from 15 to 35% on the basis of the entire quantity of waste produced. Table 2.1 summarizes the typical composition of biomedical waste. Conventionally, land and ocean dumping is considered as the major processing and disposal techniques in case of biomedical waste. However, if the wastes without disintegration are transferred and discarded in landfills, then it leads to the entry of pathogenic and disease-causing organisms into the surroundings. This, in turn, leads to the spreading of diseases in living beings. Considering the above facts, incineration and

Table 2.1 Typical composition of biomedical waste

S. no	Types of wastes	Composition by weight (%)
1	Plastic (combustible)	14
2	Dry cellublostic solid	45
3	Wet cellublostic solid	18
4	Non combustible materials	20

Source http://isebindia.com/95_99/99-07-2.html

plasma pyrolysis are found to be the best techniques for the treatment and energy recovery from biomedical wastes.

This process minimizes the quantity of biomedical wastes to 20% to 30% of its original volume. Incinerators can convert these wastes into heat, gas, steam and ash. The energy content of biomedical waste products can be exploited directly by utilizing them as a direct combustion fuel, or indirectly by recycling them into another type of fuel. The operating cost of incinerators for treating 1 kg of biomedical waste was calculated to be 0.2 USD and the energy recovered could cost around 0.12 USD from 1 kg of biomedical waste.

2.2.9 Industrial Wastes

Industrial waste is generated by industries during production, manufacturing, processing and treatment. A large number of small-scale industrial units basically deposit their waste on open land and in water bodies. This leads to serious and permanent damage to the surroundings. For example, piling of wastes from sugar manufacturing industries, food industries, paper industries, chemical processing units and pharmaceutical industries pollute the soil and water.

Industries generate various kinds of waste materials. It can exist both in solid and liquid form. The solids, especially from agro-based industries are liquefied by pretreatment methods such as thermal, chemical, physical and biological and used as a raw material for the generation of energy.

2.2.9.1 Food Processing Industries

The food processing units generate a huge quantity of waste and derivatives which could be utilized as a source of generating energy. The waste products are produced as derivatives from every division of the industries such as processing, packing and storage. Solid wastes comprise husk, leftovers of fruits, vegetables, food and coffee powders. Liquefied wastes could be produced during washing and preprocessing of meat, poultries, fisheries, fruits, vegetable processing and winemaking. The wastes generated from these industries are subjected to anaerobic digestion to generate biogas, ethanol and hydrogen.

2.2.9.2 Mash and Paper Industries

Mash and paper industries are regarded to be extremely contaminating plants and utilize a huge quantity of power and water in different parts of their functioning units. The pulp waste, paper shavings and wood wastes generated through these industries are subjected to combustion, gasification and biomethanation to generate energy. The wastewater released from these industries is extremely mixed water and it is loaded with organic substances. This organic-rich wastewater can be used to generate biogas by anaerobic digestion. Black liquor, the wastewater generated from paper industries could be prudently consumed for the generation of biogas by means of upflow anaerobic sludge blanket (UASB) technique.

2.2.9.3 Distilleries

Wastes from these industries include brewer spent grain, bagasse, residual brewing yeast, hot trub and high strength wastewater having BOD in the range of 10,000–30,000 mg/L. Solid wastes produced from distilleries undergo combustion, gasification and fermentation for the generation of ethanol and syngas. Bagasse is the fibrous matter that is left after the extraction of juice from sugarcane or sorghum during the distillation process. For every 10 tons of sugarcane crushed, a sugar factory produces nearly three tonnes of wet bagasse. It is a lignocellulosic waste that contains a good amount of raw material to support the growth of bacteria in the fermentation process. Bagasse can be used as a raw material for methane and ethanol production. Residual brewing yeast and hot trub can be used to produce animal feed and biogas. The liquid and semi-solid wastes of the distillation industry have greater potential and can be subjected to anaerobic digestion for the production of biogas.

2.2.9.4 Dairy Plants

Dairy industries produce large quantities of high-strength organic-rich wastewater in the order of magnitude of thousand cubic meters/day. With their high concentration of organic matter, the diary wastewaters create water pollution. In order to prevent water pollution, dairy industries are provided with biological wastewater treatment systems. These systems produce a huge quantity of excess sludge during the treatment of wastewater. The sludge yield for anaerobic and aerobic treatment of dairy wastewater was estimated to be 0.1 and 0.4 kg MLSS/kg COD. Sludge from wastewater treatment plant contains a good amount of biopolymers such as proteins and carbohydrates. So, it can be used as a source for the production of methane.

2.2.9.5 Sago Industries

Sago, the common edible starch in the form of globules is processed from the tubers of tapioca (*Manihotsps.*). Processing of sago generates biodegradable solid waste in the form of peeling of about 50–60 kg/ton of tuber peeled and solid fibre residues of about 55–70 kg/ton of the tuber. The waste of tapioca (*Manihotesculenta*) that is left after the processing of tuber, which is used for sago and starch production is rich in carbohydrates. Various types of microorganisms proliferate in this leftover waste. The free sugars and the sugars formed by enzymatic saccharification can be used in many biotechnological processes for single-cell protein or bioethanol production. In addition, sago industries are known for the production of biosolids from their wastewater treatment plant. Biosolids are excess microbial biomass of the biological treatment system and are frequently wasted for maintaining biomass balance inside the system. These biosolids are rich in organics and can serve as a raw material for the generation of biofuel.

2.2.9.6 Tanneries

According to conservative estimates, more than 600,000 tons per year of solid waste are generated worldwide by the leather industry, and approximately 40–50% of the hides are lost to shavings and trimmings. Tannery solid waste includes trimmings of finished leather, shaving dust, hair, fleshing and trimming of raw hides and skins. Anaerobic digestion systems are mature and proven processes that have the potential to convert tannery wastes into biogas and production of stabilized residue. This method degrades a substantial part of the organic matter in the tannery sludge and solid wastes. Upon degradation, it generate valuable biogas, which can alleviate the environmental problem. Digested solid waste is biologically stabilized and can be reused in agriculture.

The generation of energy from renewable waste is expected to cover 60% of global power capacity in the next 5 years. Among the countries, 2/3rd of its growth will be in four key markets: China (37%), the United States (13%), European union (12%) and India (9%). So it is essential to overcome the key issues associated with the generation of energy from biomass to make the process feasible. Generation of energy from waste biomass involves both upturn and clearance in the circumstance of the waste hierarchy. The benefits which waste to energy could provide and the disadvantages of the conversion process have been discussed in this chapter. This chapter discusses various key issues of waste to energy (WTE) technology. While executing a WTE technology, the key issues that need to be addressed are viability, cost of biomass feedstocks, energy potential, greenhouse gas emission, public perception and health effects.

2.3   Cost of Biomass Raw Materials

Biomass energy generation requires a feedstock that should be assembled, hauled and stored. The cost of biomass for energy generation significantly depends on the accessibility of a safe and continuous supply of suitable biomass feedstocks at a reasonable cost other than waste material. Biomass could cost around 40–50% of the total energy generation. Collection and utilization costs for biomass feedstocks are hard to estimate and no actual information is available based on costs. The cost spent for biomass will rely on its calorific value, humidity and other characteristics such as ash content, slagging and particle size.

These characteristics, in turn, will influence the cost of management or treatment at the plant and the effectiveness of energy generation. The cost calculations during biomass processing are described below.

2.3.1   Cost of Agricultural Residues

The organic waste produced as a result of agricultural activity is called agricultural residues. Agricultural residues such as paddy, wheat straw and bagasse from sugarcane are the low-cost raw materials since they can be collected during harvesting (European Climate Foundation 2010). If the agricultural residues are utilized by the owner alone, then their prices cannot be estimated clearly.

But on assessing the different parameters accountable for the entire cost of agricultural residues (including production and management), it is determined that the price of agricultural biomass may be sizeable. The total cost of agricultural residues includes: production costs, harvesting cost, collection cost, payment waged to farmers and transportation cost.

2.3.1.1   Production Cost

The production cost of agricultural residues is estimated as the product of a certain fraction of crop residues and the entire cost of crop production. This can be determined as follows [62]:

$$C_{pc} = C_{epc} * F_{ar}$$

where

C_{pc} is the production cost.

C_{epc} is the entire production.

F_{ar} is the certain fraction of agricultural residues.

2.3.1.2   Harvesting Cost

Typically, the harvesting cost is included in the production cost also. Mostly, the agricultural residues are harvested along with crops, in such cases, the harvesting cost can be estimated as follows [63]:

$$R_{pc} = C_{pc} * CR_r$$

where

R_{pc} is the residue production cost.

C_{pc} is the crop production cost.

CR_r is the crop to residue production ratio.

On the other hand, in case of some crops such as maize and cotton, the residues are harvested separately. In such circumstances, the price spent for harvesting is also taken into account for estimation. If the harvesting is carried out manually, then its cost can be calculated by dividing the daily wage rate of labour by the harvesting capacity. The harvesting cost can be estimated as follows:

$$C_{rhc} = W_{dr} / h_{ccr}$$

where

C_{rhc} is the residue harvesting cost.

W_{dr} is the daily wage rate of labour.

h_{ccr} is the harvesting capacity (amount) of crop residues of the labour/day.

2.3.1.3   Collection Cost

The residues of agricultural crops have to be assembled from the farm or agro industrial unit for piling up prior to transportation. The collection price of these residues relies on the daily wage rate and duration needed for their compilation in a specific place. The collection price can be estimated by dividing the daily wage rate by the transporting capacity (tonne per trip) and the number of transport trips [64]

$$C_{rcc} = W_r / (Ccr * n_{tt})$$

where

C_{rcc} is the residue collection cost.

W_r is the daily wage rate.

Ccr is the transporting capacity of residues.

n_{tt} is the number of transport trips of crop residues.

2.3.1.4 Transportation Cost

The collected residues are carried out from the ranch or handling unit to the utilizing point, i.e. energy converting units or to the storing place to facilitate uninterrupted supply at the time of the season. The conveyance price can be estimated as follows [64]:

$$C_{rt} = d\left(F_{co} * C_{fc} + W_{dr}\right)/(t_{cc} * t_d)$$

where

C_{rt} is the residue transportation cost.

F_{co} is the fuel consumption/h of operation.

C_{fc} is the fuel cost.

W_{dr} is the driver's wage/hour.

t_{cc} is the carrying capacity of a transportation vehicle.

t_d is the transportation distance.

2.3.1.5 Storage Cost

The storage cost holds both the handling and the capital price spent on the storage amenities. Storage cost would be the cost spent for rental of the storage area and the cost spent to wrap the residues to shield them during the rainy season. However, storage cost could not be a great impact on the cost of biomass feedstocks. Normally it is found to be negligible.

2.3.1.6 Total Cost

The total cost of agricultural residues is the summation of all costs (production cost, harvesting cost, collection cost, transportation cost and storage cost).

$$C_{tar} = C_{pc} + C_{rhc} + C_{rcc} + C_{rt}.$$

2.3.2 Cost of Forest Residues

Forest residue is a medium energy-generating biomass that competes in international markets. Feedstock supply chain effectiveness and costs are crucial while assessing the profitable sustainability of forestry residues. In case of forest residues, the cost is mainly associated with collection and conveyance. The compactness of forestry residues has a direct influence on transportation capacity. The compacted forestry residues will minimize the transportation cost substantially. The chief consideration in the forest woody biomass feedstocks are: identification of the probable woody biomass feedstocks; harvesting of woody biomass feedstocks; Stocking in the storage place; transportation to the biomass power plant; handling and managing the biomass at the plant site. Physicochemical characteristics and transportation are the two main parameters that significantly influences the cost of forest residues. The cost analysis also relies on energy cost fluctuations, labour and machine costs throughout the supply chain [63].

2.3.3 Cost of Biomass Feedstocks in Different Countries

The cost estimation for various biomass feedstocks is normally performed on the basis of their geographical origin and supply chain. Table 2.2 represents the price estimates of biomass feedstocks in different countries.

In the United States (US), sizeable amount of biomass materials are accessible from forest residuals and agricultural wastes. In case of agricultural residues, the cost spent for the residues and wastes generated from corn production is estimated to be 55 USD/tonne. It may vary from 0 to 27 USD/tonne with the approximate mean cost of about 11 USD/tonne in which a market subsist. Table 2.3 summarizes the cost estimates of biomass feedstock in different states of India.

The cost-effectiveness of these low price bagasse feedstocks makes many bioenergy power plants utilize them tremendously. In India, the price for sugarcane residue is around 12–14 USD/tonne, and the price of rice husks is around 20 USD/tonne (UNFCCC, 2011). In Punjab (India), the highest price of about 58 USD/tonne was

Table 2.2 Price estimates of biomass feedstocks in different countries

Countries	Cost of agricultural residues (USD/tonne)	Cost of forest biomass (USD/tonne)
United states	20–50	15–30
Europe	55–68	98–115
Brazil	11–13	69–71
India	22–30	56–58
China	35–52	Not available

Source Irena (2.2012)

Table 2.3 Cost of biomass feedstocks in India

Name of the states (in India)	Feedstocks cost in USD/tonne	
	Bagasses	Biomass
Maharashtra	57	39
Haryana	56	39
Uttar Pradesh	50	31
Andhra Pradesh	49	28
Rajasthan	49	34
Tamil Nadu	48	30
Other states	52	34

Source Central Europe Regional Contest, CERC (2.2015)

spent on biomass feedstocks, whereas a low cost of about 48 USD/tonne was spent in Tamil Nadu, India. In other states, a cost of about 50 USD/tonne was incurred for biomass feedstocks.

2.3.4 Cost of Energy and Bioelectricity

Waste to energy (WTE) technologies generate energy in the form of heat or electricity in a combined hear and power generator unit. The produced energy can be utilized in several ways. The source of energy supply is also an essential factor that could be taken into consideration in bioelectricity generation. Utilization of energy from residual waste prior to ultimate disposal could create a reasonable and major sustainable contribution to the energy strategy. Energy and power generation from residual derived fuels through waste to energy conversion technologies affords energy and power at balanced costs for industries. Energy from waste biomass can replace other modes of energy-intensive power generation. The estimation of energy produced by various conversion processes is listed in Table 2.4.

Table 2.4 Estimation of energy produced by various conversion processes

S. no	Bioenergy	Conversion process	Cost of energy ($/kWh)	References
1	Heat	Combustion	2.1–2.7	[65]
2	Syngas	Gasification	2.02–2.56	[65]
3	Bio oil	Pyrolysis	1.91	[66]
4	Biogas	Anaerobic digestion	0.08	[67]
5	Bioethanol	Fermentation	0.1	[67]

2.4   Climate Changes and Greenhouse Gas Emissions

Waste to energy leads to the minimization of greenhouse gas emissions. Figure 2.1 explains how waste to energy technologies helps in controlling climate change and greenhouse gas emissions. Biomass waste coming out from the manufacturing process (paper industries) can be reused or used to generate energy through waste to energy technology. This can minimize the usage of fossil fuels and increase carbon sequestration. Solid waste management is an industry incorporating WTE technology that can reduce climate change and greenhouse gas emissions in the following three ways:

- Combustion of biomass can reduce the usage of fossil fuels. There are over 800 of these plants worldwide and are used to produce electrical energy and heat energy for the community
- Biodegradable waste can be composted into value-added fertilizers. As a result, carbon is stored in soil and will be utilized by plants.
- Landfilling of biomass resulted in the production of fuel gas, methane, which can reduce or replace the usage of fossil fuels. However, in most cases, it is also

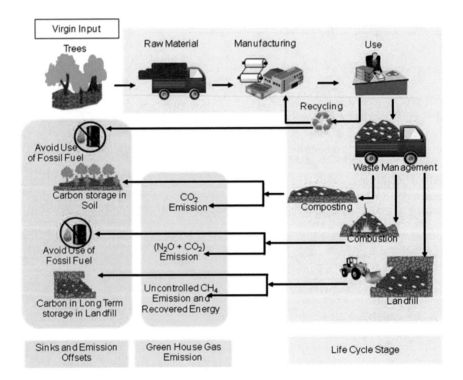

Fig. 2.1 WTE technologies in controlling climate change and green house gas emissions

responsible for the addition of greenhouse gases to the atmosphere. Long-term storage of carbon in landfills increases soil fertility.

Composting is the most preferable method for managing organic waste than landfilling, where it is impossible to control the fugitive emission of methane. Methane is a greenhouse gas and its impact is more than 20 times that of carbon dioxide. Greenhouse gas-generating potential of aerobic composting is comparatively very less than landfilling, where it generates methane, a powerful greenhouse gas. In addition, the use of compost reduces the watering of plants up to 70% and minimizes the usage of chemical fertilizers which are said to be energy-intensive.

2.5 Public Perception

Recent advances in alternative energy dramatically improved technologies responsible for biofuel production in a feasible manner. However, the public is having different perceptions about the recovery of energy from waste biomass. So it is very important to consider their views to overcome the barriers associated with the biomass bioenergy economy. Consumers are public and if their perception about the purchase and consumption of bioenergy is negative, then the market for bioenergy will get disturbed. Therefore, it is essential to understand public perception of continued biofuel development.

2.5.1 Sources of Information for Public Perception

Awareness about bioenergy through media such as TV, internet and blogs is often beneficial. Sources of information for the public include internet, NGOs, government policies, politicians, environmental campaign and media. Figure 2.2 represents the various sources of information that influence the public.

Among these, media plays a significant role in changing public perception about bioenergy from biomass. Most of the time it is negative, but there are some positive incidents also. Positive bioenergy information from government departments is often not trusted by the people. Information sources from NGOs influence public perception at a greater extent, as they are part of the community. Normally NGOs either stand for or against bioenergy projects.

The other sources of information from internet, politicians and environmental awareness campaign can significantly influence the people mind set. Due to the influence of these information sources, the public either positively or negatively argues about biomass-based bioenergy. Positive arguments include: renewable energy, development of village, recycling of biomass, eco-friendly job, sustainable forestry and energy security. Negative arguments include: carbon emission, introduction of

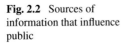

Fig. 2.2 Sources of information that influence public

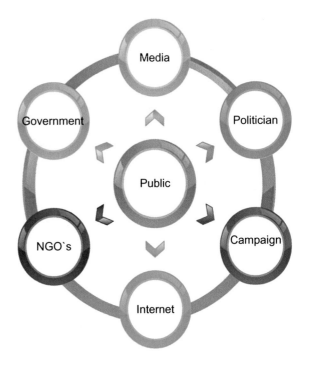

non-native plants, fragmentation of agricultural land and associated environmental pollution.

Limited public understanding and false information often confuse the public and it increases the risk of implementing bioenergy-based projects. Hence, it is important for the policymakers to find reliable information sources to spread real scenarios about biomass bioenergy projects. In addition, frequent education campaigns focusing on positive and negative impacts should be conducted regularly to cease fear among the public. Companies can use videos, fact sheets and research findings to spread information.

2.5.2 Health Issues

The conventional biomass utilization as energy leads to indoor air pollution and emission of greenhouse gases which are considered to be serious health hazards. The health and ecological effects of conventional modes of biomass disposal on land are methane (a greenhouse gas) emission and leachate formation. Incineration of waste leads to continuous release of dust, metals, acids, dioxins and other contaminants into environment. Waste to energy technologies (WTE) such as biogas generation, ethanol production, syngas fermentation and gasification for power generation can reduce air pollution, minimize the emission of greenhouse gases and reduces land degradation and land reclamation. Good operating practices under controlled conditions can be adopted in the incineration of waste biomass to generate thermal energy and electricity. In the same way, the anaerobic digestion process can be used to produce methane from municipal solid waste for cooking and electricity production. These technologies reduce air pollution, minimize greenhouse gas release and reduce waste disposal through suitable waste managing practices.

2.5.3 Dust/Odour

During the waste management process, dirt and odours are produced continuously. This can be reduced through a better construction plan, doing all operations in controlled conditions inside the system, fine operation and effectual handling are carried out for dirt control by means of transport systems. The elimination of odour from waste treatment plants or incineration amenities requires cautious contemplation. As waste incineration plant is usually within roofed constructions, probable smell releases may usually be processed via the efficient aeration process.

2.5.4 Flies, Vermin and Birds

The closed system of waste incineration processing units would restrict the potential to focus on vermin and birds. On the other hand, in an open system, in a hot climate, it is probable that the flies to be gathered, particularly when they accumulate at the time of disposal of waste. Efficient maintenance and on-spot managing of tilting and storing places are mandatory to reduce the menace from vermin and insects. In certain cases, the old waste should be properly blended with fresh waste, so the temperature surpasses the point at which flies may live. Likewise, carefully maintaining the period of waste storage in such a way that it should not exceed the reproduction cycle of vermin for instance rats.

2.5.5 Noise

Noise is a problem that can be managed through authorizing policy. There is a limitation for noise extent by perceptive receptors which could be suggested and recommended by a scheduling authorization. The major sources of noise during handling and processing include the following:

- Vehicle transport/strategy;
- Mechanical treatments such as waste grounding;
- Air withdrawal and aeration processes;

- Vapour turbine plants; and
- Air chilled/condensor system.

2.5.6   Water

The enclosed structure of operation in waste managing processes extensively minimizes the eventual influence and impact on the water surroundings. Contamination of water bodies is mainly due to leachate originating from biomass feedstocks. These, in turn, result in seepage and contaminate the water bodies. Therefore, a management design is needed to overcome the problem of seepage. Agricultural-based bioenergy potentially impacts water resources in arid and semi-arid regions. Excessive use of water for biomass production and feedstock conversion in these area often leads to water shortage and groundwater depletion.

References

1. Ronak agrotek (2015) Calorific value chart. Class 7 under no 16423425
2. Echiegu EA, Nwoke OA, Ugwuishiwu BO, Opara IN (2013) Calorific value of manure from some nigerian livestock and poultry as affected by age. Int J Sci Eng Res 4:999–1004
3. Su P, Brookes PC, He Y, Wu J, Xu J (2016) An evaluation of a microbial inoculum in promoting organic C decomposition in a paddy soil following straw incorporation. J Soil Sediment 16(6):1776–1786
4. Hiloidhari M, Baruah DC (2011) Rice straw residue biomass potential for decentralized electricity generation: a GIS based study in Lakhimpur district of Assam, India. Energy Sust Develop 15:214–222
5. Madhuri N, Velmurugan B, Anil K, Bhim SP, Murari S (2016) Enhanced biogas production from rice straw by selective micronutrients under solid state anaerobic digestion. Bioresour Technol 220:666–671
6. Balachandra P (2011) Modern energy access to all in rural India: An integrated implementation strategy. Energy Policy 39:7803–7814
7. Arry YN, Yuda CH, Hasanah W (2016) Endeavoring to food sustainability by promoting corn cob and rice husk briquetting to fuel energy for small scale industries and household communities. Agric Agric Sci Procedia 9:386–395
8. Balat M (2011) Production of bioethanol from lignocellulosic materials via the biochemical pathway: a review. Energy Convers Manage 52(2):858–875
9. Omidvar M, Karimi K, Mohammadi M (2016) Enhanced ethanol and glucosamine production from rice husk by NAOH pretreatment and fermentation by fungus Mucorhiemalis. Biofuel Res J 3(3):475–481
10. Buranov AU, Mazza G (2008) Lignin in straw of herbaceous crops. Ind Crops Prod 28:237–259
11. Jelle W, Arjan TS, Johannes HR, Wouter JJH (2013) Ethanol-based organosolv fractionation of wheat straw for the production of lignin and enzymatically digestible cellulose. Bioresour Technol 135:58–66
12. Mandar PB, Parag RG, Aniruddha BP, Levente C (2014) Hydrodynamic cavitation as a novel approach for delignification of wheat straw for paper manufacturing. Ultrason Sonochem 21(1):162–168

13. Meena K, Virendra KV, Ram C (2016) Performance evaluation of various bioreactors for methane fermentation of pretreated wheat straw with cattle. Green Proc Synthesis 5(2):113–121. https://doi.org/10.1515/gps-2015-0067

14. Arora D, Sharma V, Garg U, Kumar (2016) Decolourizing of distillery spent wash using indigenously prepared cation exchanger from the agricultural waste (Wheat Straw). Int J Emerg Techno. 7.1:11–17

15. Tsai WT, Chang CY, Wang SY, Chang CF, Chien SF, Sun HF (2001) Utilization of agricultural waste corn cob for the preparation of carbon adsorbent. J Environ Sci Health B 36(5):677–686. https://doi.org/10.1081/PFC-100106194

16. Lin L, van der Ester V, Gjalt H (2009) An energy analysis of ethanol from cellulosic feedstock Corn stover. Renew Sust Energy Rev 13:2003–2011

17. Wang X, Feng Y, Wang H, Qu Y, Yu Y, Ren N, Li N, Wang E, Lee H, Bruce EL (2009) Bioaugmentation for electricity generation from corn stover biomass using microbial fuel cells. Environ Sci Technol 43(15):6088–6093

18. Dotaniya ML, Datta SC, Biswas DR, Dotaniya CK, Meena BL, Rajendiran S, Regar KL, Manju L (2016) Use of sugarcane industrial by-products for improving sugarcane productivity and soil health. Int J Recycl Organ Waste Agric 5(3):185–194

19. Agrawal KM, Barve BR, Khan SS (2013) Biogas from press mud. IOSR J Mech Civil Eng 37–41

20. Bahurudeen A, Marckson AV, Kishore A, Santhanam M (2014) Development of sugarcane bagasse ash based Portland pozzolana cement and evaluation of compatibility with super plasticizers. Constr Build Mater 68:465–475

21. Sessa TDC, Silvoso MM, Vazquez EG, Qualharini EL, Haddad AN, AmaralAlves L (2016) Study of the technical capability of sugarcane bagasse ash in concreteproduction. Mater Sci Forum 866:53–57

22. Azma P, Yasseer A, Hady E, Wan MF, MdRazali A, Muhammad SP (2013) Utilizing sugarcane wasted fibers as a sustainable acoustic absorber. Procedia Eng 53:632–638

23. Manoj T, Sahub JN, Ganesan P (2016) Effect of process parameters on production of biochar from biomass waste through pyrolysis: a review. Renew Sustain Energy Rev 55:467–481

24. De Vrieze J, Raport L, Roume H, Vilchez-Vargas R, Jáuregui R, Pieper DH, Boon N (2016) The full-scale anaerobic digestion microbiome is represented by specific marker populations. Water Res 104:101–110

25. Wang P, Yu Z, Zhao J, Zhang H (2018) Do microbial communities in an anaerobic bioreactor change with continuous feeding sludge into a full-scale anaerobic digestion system? Bioresour Technol 249:89–98

26. Lee J, Shin SG, Han G, Koo T, Hwang S (2017) Bacteria and archaea communities in full-scale thermophilic and mesophilic anaerobic digesters treating food wastewater: Key process parameters and microbial indicators of process instability. Bioresour Technol 245:689–697

27. Lin L, Yu Z, Li Y (2017) Sequential batch thermophilic solid-state anaerobic digestion of lignocellulosic biomass via recirculating digestate as inoculum—Part II: Microbial diversity and succession. Bioresour Technol 241:1027–1035

28. Ge X, Xu F, Li Y (2016) Solid-state anaerobic digestion of lignocellulosic biomass: recent progress and perspectives. Bioresour Technol 205:239–249

29. Balan V (2014) Current challenges in commercially producing biofuels from lignocellulosic biomass. ISRN Biotechnol 2014:1–31

30. Raposo F, De la Rubia MA, Fernández-Cegrí V, Borja R (2012) Anaerobic digestion of solid organic substrates in batch mode: an overview relating to methane yields and experimental procedures. Renew Sustain Energy Rev 16:861–877

31. Brown D, Li Y (2013) Solid state anaerobic co-digestion of yard waste and food waste for biogas production. Bioresour Technol 127:275–280

32. Angelidaki I, Sanders W (2004) Assessment of the anaerobic biodegradability of macropollutants. Rev Environ Sci Biotechnol 3:117–129

33. Xu F, Li Y, Ge X, Yang L, Li Y (2018) Anaerobic digestion of food waste—challenges and opportunities. Bioresour Technol 247:1047–1058

34. Cerda A, Artola A, Font X, Barrena R, Gea T, Sánchez A (2018) Composting of food wastes: status and challenges. Bioresour Technol 248:57–67
35. Wang X, Selvam A, Lau SSS, Wong JWC (2018) Influence of lime and struvite on microbial community succession and odour emission during food waste composting. Bioresour Technol 247:652–659
36. Zhang H, Schroder J (2014) Animal manure production and utilization in the US. Appl Manure Nutr Chem Sustain Agric Environ 1–21
37. Kumar S (2010) Composting of municipal solid waste. Crit Rev Biotechnol 31:112–136
38. Li YF, Nelson MC, Chen PH, Graf J, Li Y, Yu Z (2014) Comparison of the microbial communities in solid-state anaerobic digestion (SS-AD) reactors operated at mesophilic and thermophilic temperatures. Appl Microbiol Biotechnol 99:969–980
39. De Baere L, Mattheeuws B (2012) Anaerobic digestion of the organic fraction of municipal solid waste in Europe. Waste Manag 3:517–526
40. Mehta CM, Palni U, Franke-Whittle IH, Sharma AK (2014) Compost: Its role, mechanism and impact on reducing soil-borne plant diseases. Waste Manag 34:607–622
41. Jurado M, López MJ, Suárez-Estrella F, Vargas-García MC, López-González JA, Moreno J (2014) Exploiting composting biodiversity: Study of the persistent and biotechnologically relevant microorganisms from lignocellulose-based composting. Bioresour Technol 162:283–293
42. Liang Y, Lu Y, Li Q (2016) Comparative study on the performances and bacterial diversity from anaerobic digestion and aerobic composting in treating solid organic wastes. Waste Biomass Valor 8:425–432
43. Sharmila VG, Angappane S, Gunasekaran M, Kumar G, Banu JR (2020) Immobilized ZnO nano film impelled bacterial disintegration of dairy sludge to enrich anaerobic digestion for profitable bioenergy production: Energetic and economic analysis. Bioresour Technol 308:123276
44. Aboudi K, Álvarez-Gallego CJ, Romero-García LI (2015) Semi-continuous anaerobic co-digestion of sugar beet byproduct and pig manure: effect of the organic loading rate (OLR) on process performance. Bioresour Technol 194:283–290
45. Sharmila VG, Banu JR, Kim S-H, Kumar G (2020) A review on evaluation of applied pretreatment methods of wastewater towards sustainable H_2 generation: energy efficiency analysis. Int J Hydrogen Energy 45:8329–8345
46. Uthirakrishnan U, Sharmila VG, Merrylin J, Kumar SA, Dharmadhas JS, Varjani S, Banu JR (2022) Current advances and future outlook on pretreatment techniques to enhance biosolids disintegration and anaerobic digestion: a critical review. Chemosphere 288:132553
47. El-Mashad HM, Zhang R (2010) Biogas production from co-digestion of dairy manure and food waste. Bioresour Technol 101:4021–4028
48. Banu JR, Sharmila VG, Kannah RY, Kanimozhi R, Elfasakhany A, Gunasekaran M, Kumar SA, Kumar G (2022) Impact of novel deflocculant ZnO/Chitosan nanocomposite film in disperser pretreatment enhancing energy efficient anaerobic digestion: parameter assessment and cost exploration. Chemosphere 286:131835
49. Chen H, Yan S-H, Ye Z-L, Meng H-J, Zhu Y-G (2012) Utilization of urban sewage sludge: chinese perspectives. Environ Sci Pollut Res 19:1454–1463
50. Zhang J, Lü F, Shao L, He P (2014) The use of biochar-amended composting to improve the humification and degradation of sewage sludge. Bioresour Technol 168:252–258
51. Sokkanathan G, Sharmila VG, Kaliappan S, Banu JR, Yeom IT, Rani RU (2018) Combinative treatment of phenol-rich retting-pond wastewater by a hybrid upflow anaerobic sludge blanket reactor and solar photofenton process. J Environ Manage 206:999–1006
52. Chang H-D, Chen C-Y (2014) Composting of biosolids enhanced by a combined pretreatment with hydrogen peroxide and triton X-100. Waste Biomass Valor 6:45–51
53. Fu B, Jiang Q, Liu H-B, Liu H (2015) Quantification of viable but nonculturable *Salmonella* spp. and *Shigella* spp. during sludge anaerobic digestion and their reactivation during cake storage. J Appl Microbiol 119:1138–1147
54. Fu B, Jiang Q, Liu H, Liu H (2013) Occurrence and reactivation of viable but non-culturable *E. coli* in sewage sludge after mesophilic and thermophilic anaerobic digestion. Biotechnol Lett 36:273–279

55. Sharmila VG, Kavitha S, Obulisamy PK, Banu JR (2020) Production of fine chemicals from food wastes. In: Food waste to valuable resources. Elsevier, pp 163–188

56. Chynoweth DP, Turick CE, Owens JM, Jerger DE, Peck MW (1993) Biochemical methane potential of biomass and waste feedstocks. Biomass Bioenerg 5:95–111

57. Zhang L, Sun X (2014) Changes in physical, chemical, and microbiological properties during the two-stage co-composting of green waste with spent mushroom compost and biochar. Bioresour Technol 171:274–284

58. Martin-Ryals A, Schideman L, Li P, Wilkinson H, Wagner R (2015) Improving anaerobic digestion of a cellulosic waste via routine bioaugmentation with cellulolytic microorganisms. Bioresour Technol 189:62–70

59. Cirne DG, Björnsson L, Alves M, Mattiasson B (2006) Effects of bioaugmentation by an anaerobic lipolytic bacterium on anaerobic digestion of lipid-rich waste. J Chem Technol Biotechnol 81:1745–1752

60. Sharmila VG, Kumar SA, Banu JR, Yeom IT, Saratale GD (2019) Feasibility analysis of homogenizer coupled solar photo Fenton process for waste activated sludge reduction. J Environ Manage 238:251–256

61. Nakasaki K, Mimoto H, Tran QNM, Oinuma A (2015) Composting of food waste subjected to hydrothermal pretreatment and inoculated with *Paecilomyces* sp. FA13. Bioresour Technol 180:40–46

62. Tripathi AK, Iyer PVR, Kandpal TC, Singh KK (1998) Assessment of availability and costs of some agricultural residues used as feedstocks for biomass gasification and briquetting in india. Energy Convers Mgmt 39:1611–1618

63. Fagernas L, Johansson A, Wilén C, Sipilä K, Mäkinen T, Helynen S, Daugherty E, den Uil H,Vehlow J, Kåberger T, Rogulska M (2006) Bioenergy in Europe. Opportunities and barriers. Research Notes 2352. Espoo. 118

64. Tidball R, Bluestein J, Rodroguez S, Knoke R (2010) Cost and performance assumptions for modeling electricity generation technologies, NREL, USA

65. Tyrol (2000) Contributions ECN biomass to "Developments in thermochemical biomass conversion" Conference, Austria, ECN-RX-00-026. 17–22

66. Rogers JG, Brammer JG (2012) Estimation of the production cost of fast pyrolysis bio-oil. Biomass Bioenerg 36:208–217

67. Mohan N, Paroha S, Kumar S (2016) Feasibility of ethanol production in India through alternate feed stocks. Presented during ISSCT XXIX Congress held from 5th to 8th December- 2016 at Thailand

Chapter 3
Enhancement of Hydrolysis

The relationship between structural components of carbohydrate for a given biomass type reflects its complexity. Crystallinity and degree of polymerization of cellulose, the presence of an extracellular layer (EPS layer) in the sludge biomass, available surface area (or penetrability). lignin fortification. cellulose sheathing through hemicellulose and fibre strength are the factors that lead to biomass recalcitrance [1]. This recalcitrance is responsible for variations in the hydrolysis/digestibility of a specific biomass feedstock. The lignin removal and EPS extraction improve biomass digestibility until the impact of the residual lignin and EPS is no more satisfactory to impede microbial digestion or enzymatic hydrolysis [2]. When the effect of lignin is minimized, an extreme crystalline structure of cellulose that makes it less approachable to cellulase destruction than amorphous cellulose in which cellulose accessibility to cellulase is found to be the most significant rate restrictive aspect in enzymatic hydrolysis [3]. Upstream activities such as reduction in size and physicochemical pretreatment should indeed be prioritized in bio-based exploration attempts directing woody biomass for optimizing the enzymatic hydrolysis, direct microbial utilization of polysaccharides and sugar yields [4]. Certain pretreatment technologies are hard because it incorporates upstream and downstream processing costs, capital expenditure, chemical recycling and waste treatment systems which are difficult to analyze and compare [5]. Mass balance analysis, on the other hand, can be used to assess the pretreatment efficacy of a process for a given feedstock as a part of an industrial system or biorefinery [6]. This needs a detailed commercial investigation to find the optimal pretreatment method determining the specific local feedstock in the industrial process, particularly in the case of co-location of present plants with steam, low-cost power, or default treatment is obtainable.

Since these pretreatments can effectively solubilize organic matter in sludge, they can be improved by extracting the EPS layer. EPS is a cell layer capable of producing flocs by aggregating microbial cells and obstructing microbial cell accessibility during pretreatment. Protein, carbohydrates, nucleic acid and lipids constitute EPS [7]. It is divided into two types: loosely bound and tightly bound EPS. EPS extraction usually involves separating microbial cells by removing their adhesion

© Springer Nature Singapore Pte Ltd. 2022
K. Sudalyandi and R. Jeyakumar, *Biofuel Production Using Anaerobic Digestion*,
Green Energy and Technology, https://doi.org/10.1007/978-981-19-3743-9_3

[8]. The pretreatment process is improved by extracting the EPS and liberating the adsorbed organic biopolymers in the extracellular layer.

3.1 Trends in Liquefaction (Combinative and Others)

Liquefaction is defined as the structural breakdown of organic materials into an aqueous form. The microbial cell that exists in sewage sludge biomass consists of a complicated lattice structure [9–12]. This structure hinders the substrate accessibility during the hydrolysis process affecting biodegradation. This urges the proper pretreatment of sludge before the digestion process for rectifying these issues [13]. The Lignocellulosic biomass (energy crops and plant waste) consists of cellulose, hemicellulose and lignin in various proportions [14–17]. Lignin is the recalcitrant compound that exists in this biomass and makes it insoluble in water and microbially resistant. Hence, proper pretreatment is necessary to enhance biodegradability during the AD process [18–26]. Also, manure appears to be amenable to pretreatment, although the properties of the raw substrate as well as with other substrates influence the pretreatment outcome [27–37]. The properties of manure vary based on the type, e.g. cow dung being more resistant than pig manure. Due to the obvious loss of organic material in the form of lipids, proteins and VFA, as well as the generation of inhibitory compounds, pretreatment may reduce the biodegradability of pig dung [38]. When pretreatment was applied to cow dung containing less readily degradable chemicals, the COD solubilization resulted in improved biodegradability with no signs of inhibition. Pre-treatments have comparable effects on animal manure as that of fibrous plant material, namely, the solubilization of lignin and hemicelluloses, which is caused by all pretreatments except mechanical. Ultrasonic and other mechanical pretreatments have produced particle size reduction [39–41]. Pretreatment of food waste has a wide range of effects that are dependent highly on the pretreatment process and the substrate constitution. Figure 3.1 lists the various types of pretreatments. Organic waste from households (Organic Fraction of Municipal Solid Waste—OFMSW) consists of complex components that are highly biodegradable and depend largely on upstream processing to remove unwanted fractions and homogenize the remaining material.

The vital role of the sludge pretreatment process is to cleave the cell wall of microbes that enhance organic solid reduction and promotes methane generation. Usually, chemical pretreatment in conjunction with high temperature is highly preferred for lignocellulosic biomass. In addition, mechanical, thermal and wet oxidation (WO) pretreatment also enhances the lignocellulosic biomass degradability. All these pretreatments solubilize the hemicellulose and lignin which exposes the cellulose for digestion. Thus, promoting the surface area accessibility by removing the chemical composition of lignin which contributes to biodegradability improvement.

Some of the pretreatment processes such as ultrasonic, microwave and homogenizers consume more input energy, but the generation of output energy in the form of

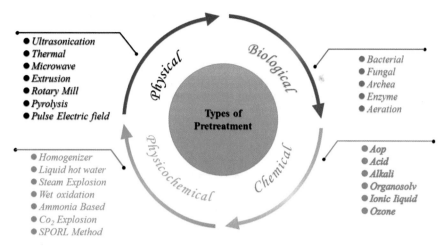

Fig. 3.1 Classification of pretreatment techniques

methane is high [11]. The ultrasonic pretreatment of sludge prior to anaerobic digestion increased biodegradability which becomes more soluble by dismantling the cell
structures of bacteria and reducing the particle size of fat globules. The ultrasonic
pretreatment of sludge enhanced 104% of the methane from 143 to 292 CH_4/kg DS
during digestion which was operated at 20 kHz and 0.33 W/mL for 20 min. Though
the ultrasonic pretreatment consumes more specific energy, it reduces the digestion
period thereby minimizing the power consumption in digestion [11, 12]. About 20
full-scale and 77 pilot-scale installations of ultrasonic sludge pretreatment are available, but the majority of them are in Germany. This pretreatment enhanced 50–35%
of increased methane output with significant solid elimination from waste biomass.
The ultrasonic pretreatment of WAS diminishes the floc size in the reduced energy
input of 1000 kJ/kg TS. In higher energy input of 2000 kJ/kg TS, the floc disintegration occurs resulting in the expulsion of extracellular or intracellular organic
components. With higher and strong enough ultrasonic input power, 100% disintegration can be achieved. Lower energy inputs for ultrasonic pretreatment of pig
manure have also been proposed as the most efficient, resulting in solubilization and
particle size reduction, as well as increased biodegradability. This substrate was also
shown to be more amenable to ultrasonic treatment than WAS, since comparable
COD solubilization was obtained with lower specific energy inputs.

Similarly, microwave and thermal pretreatment of sludge also have greater effectiveness in sludge disintegration before AD [13]. The microwave pretreatment has
an effective sludge liquefaction rate enhancing 50% bioenergy generation. The food
industrial sewage sludge undergone MW pretreatment enhanced 3.1-fold adsorption
and 1.7-fold biogas output. In addition, the solubilization efficiency of sludge gets
improved from 2% (control) to 22% during this pretreatment. The biological acidogenic process improved from 3.58 to 4.77 g COD/L. (2450 MHz, 700 W for 15 min)
[14]. It was found a wide range of impacts on energy in crops and plant residues

during thermal treatment. The temperature increment of about 150–180 °C causes hemicellulose and lignin solubilization by thermal effect.

Acidic and alkali pretreatment has multiple advantages with the solubilization of biomass holding a lot of promise. It makes use of a simple device and is easy to operate which high methane conversion efficiency and low-cost consumption [15]. Acids such as HCl, H_2SO_4, H_3PO_4 and HNO_3 were used in acidic pretreatment, and alkalies such as NaOH, KOH, Ca(OH)$_2$, Mg(OH)$_2$, CaO and ammonia have been used in alkaline pretreatment. The acid/alkali pretreatment avoids the requirement of high temperature; thus, it can be used at room temperature or at moderate temperatures. This pretreatment efficiency varies based on the types and characteristics of substrates having organic components with a strong affinity [16].

AOP is a highly attractive and promising method in the sludge management process. In this process, liquefaction of sludge is carried out by highly reactive oxidizing species hydroxyl radicals [17, 18]. This radical promotes cell wall oxidation and enhances the discharge of intracellular cellular substances. Various types of methods are available in advanced oxidation pretreatments such as Fenton pretreatment, Photocatalytic pretreatment, H_2O_2 oxidative pretreatment and its combination. Of these pretreatments, Fenton pretreatment showed better efficiency and was highly relative to the other pretreatment process [18–20]. In the Fenton method, hydrogen peroxide (H_2O_2) and catalyst iron ions (Fe^{2+}) interacted for generating the highly potent hydroxyl radicals (OH) to induce oxidation of organic substances. Hydroxyl radicals have high redox potential than hydrogen peroxide (+1.36 V) and ozone (+2.07 V) [22]. This generated OH radicals dissociate the EPS of the sludge, and results in the emancipation of both intracellular substances and bound water. This instance stimulates the degradation process [22, 23]. The given equations elucidate the oxidation of organic compounds (RH) during Fenton's process.

$$Fe^{2+} + H_2O_2 \rightarrow Fe^{3+} + OH^- + OH \tag{3.1}$$

$$Fe^{2+} + \cdot OH \rightarrow Fe^{3+} + OH^- \tag{3.2}$$

$$RH + \cdot OH \rightarrow H_2O + R \tag{3.3}$$

$$R \cdot + Fe^{3+} \rightarrow R + Fe^{2+} \tag{3.4}$$

This process oxidizes the organic substances within a fraction of seconds but had the drawback of Fenton sludge formation. Even though this method showed a strong base, additional development was necessary and yet to be explored. Wet oxidation (WO) pretreatment was implemented for the sludge mixture of 2:1 (WAS:PS). This pretreatment promotes COD solubilization in which most of the solubilized organic gets oxidized to CO_2 [25–28]. This results in the existence of a low quantity of organic substance for AD NO^- is optimized that will enhance AD [29].

Usually, mechanical pretreatment is preferred for upstream processing of OFMSW. In addition to mechanical pretreatments [30–46], pretreatments such as microwave, PEF, freeze/thaw, and WO were also used [47, 48]. Mechanical pretreatments are the sole way to reduce the particle size of OFMSW and all other pretreatments have been demanded to improve biodegradability. Based on the particle size of OFMSW, the pretreatment procedure varies and has a satisfactory impact on the solubilization and reduction efficiency. The kitchen trash is exposed to five different pretreatments namely: chemical (acid to pH 2), thermal (120 °C, 1 bar), thermochemical (120 °C, pH 2), mechanical (pressure 10 bar) and freeze–thaw (−80 °C). All pretreatments resulted in COD solubilization, however, thermochemical has the most effective solubilization and mechanical has the least effective [49, 50]. However, different mechanisms of solubilization resulted in varying degrees of biodegradability; mechanical solubilization resulted in the highest degree of biodegradability, whereas solubilization caused by acid addition was likely accompanied by the formation of inhibitory/refractory compounds, which hampered biodegradability [51]. Microwave pretreatment (175 °C) of model kitchen waste efficiently solubilized proteins and sugars, but biodegradability enhancement did not always follow the solubilization pattern which depends on the heating rate. The application of WO had a negligible effect on the biodegradability of food waste.

The majority of pretreated organic wastes from the food sector are slaughterhouse waste or waste from the dairy industry. Thermal and chemical pretreatments were the most common, followed by ultrasonic and microwave pretreatments [52]. Chemical and ultrasonic pretreatments are the only ones that cause particle size reduction in food waste, although all pretreatment methods result in solidification [53]. The effects on biodegradability vary depending on the raw substrate characteristics and might range from a decrease to an increase for most pretreatments [54]. Several portions of slaughterhouse waste hold high primary biodegradability, which results in no impact on pretreatment, as seen with thermal (70 and 133 °C) and chemical (alkali) pretreatments [55]. Chemical (acid and alkali), as well as ultrasonic and low-temperature thermal (70 °C) pretreatment of various meat processing substrates, resulted in a significant increase in soluble COD [56]. Both alkali and acid addition effectively solubilized protein-rich materials and carbohydrates, however, it is less efficient at solubilizing lipids, but ultrasonic was the most effective pretreatment [57]. In a few situations, these solubilizations resulted in increased biodegradability, but biodegradability was reduced in the majority of cases due to inhibitory product production. The thermal processing of slaughterhouse waste at high temperatures (133 °C, >3 bar) has also been linked to the formation of hazardous chemicals [58].

Feathers are the byproducts of the slaughterhouse waste and are primarily consist of keratin, a recalcitrant substance. The application of chemical (alkaline) and thermal (70 and 120 °C) pretreatments to feathers improve the biodegradability of keratin during anaerobic digestion [59]. Various streams of waste from the dairy processing industry have varying initial biodegradability. Various combined pretreatment has been adopted to assess the potency of better pretreatment methods such as lower energy consumption and substrate properties. One such effort is the thermal pretreatment of plant residues which has effective COD solubilization at 121 °C followed by mechanical pretreatment. This may substantially consume lower energy than the pulse electric field (PEF) pretreatment of WAS [24]. On comparing MW and thermal pretreatment around the same temperature, the MW pretreatment contributes to a high COD solubilization for primary sludge (PS) and waste activated sludge (WAS) at 65 °C. Both treatments utilize comparable energy requirement [25]. On comparing ultrasonication with low thermal pretreatment of WAS at 80–90 °C [26]. The liquefaction of proteins and carbohydrates occur for microbial cells with/without EPS which are abundant in carbohydrates and are solubilized up to the temperature of 200 °C. The solubilization process varies linearly. At temperature above 165 °C, the COD becomes recalcitrant [27]. Temperature over 135 °C increases COD solubilization and biodegradability of the animal manure. Temperature in excess of 165 °C generate recalcitrant organic substances. The impact may be amplified by adding alkali and biodegradability worsened at pH 12, implying the formation of inhibitory chemicals [60, 61]. Increasing the temperature over 100 °C did not enhance biodegradability anymore, and the same effect could be accomplished at a lower temperature of 70 °C by adding alkali to pH 14.

All pretreatments, with the exception of low-temperature thermal and microwave pretreatments, might improve biodegradability [60]. Furthermore, low-temperature thermal and microwave pretreatments have negative consequences such as formation of refractory compounds and loss of organic matter due to volatilization. Most of the pretreatments improve biodegradability but a little or negative impact has been recorded for rare cases. Initially, mechanical pretreatment is carried out for particle size reduction. In most of the cases, high temperature thermal, MW, and all types of chemical pretreatment, refractory compounds gets developed and the organic material is lost during WO and alkali pretreatment [44]. Acid or alkali addition also improves the liquefaction of lignin. Chemicals, as well as thermal pretreatments, may generate recalcitrant chemicals from lignin, mainly phenolic compounds, which begin to develop at about 160 °C of temperature. This generated phenolic compounds act as an inhibitor for methanogens affecting the AD system [39]. The WO pretreatment converts lignin into biodegradable compounds. Non-selective oxidative pretreatments might degrade the solubilized organic material for AD [40]. Mechanical pretreatment lowers particle size and does not create inhibitors. Substrates with high fibre content are usually treated by cutting and stirring ball mills which result in particle size reduction and COD solubilization, thereby improving biodegradability [41, 45].

The physical pretreatment such as ultrasonic, high and low-temperature thermal and microwave pretreatment has only a linear correlation between biodegradability and organic solubilization [30, 31]. In case of low-temperature thermal pretreatment (70 °C), the solubilization of VS could be seen. The ultrasonic pretreatment exhibited better particle size reduction and solubilization caused by heat generated during the process where the temperature rises up to 170–190 °C for the specific energy of 6250–9350 kJ/kg TS [32]. But Ozone pretreatment has a low impact compared to ultrasonication. Similarly, freeze/Thaw pretreatment of mixed sludge increased the solubilization of COD [33]. But inhigh-temperature pretreatments above 170 °C causes partial loss of soluble organic release. The paper and pulp industrial sludge have low biodegradability during AD on pretreating the waste before digestion which enhances the biodegradability by influencing the sludge properties [34, 35]. Based on the industrial sludge age and sludge biodegradability, pretreatment impact varies for all PS, WAS and mixed sludge. Sludge with low biodegradability requires more impact to reflect better solubilization [36, 62–66].

3.2 Phase Separation

Even though pretreatment could effectively solubilize the organic matter in sludge, its efficiency can be improved by extracting the EPS layer. EPS are the cell layer capable of producing flocs by aggregating microbial cells hindering the accessibility of microbial cells during pretreatment [67, 68]. EPS are made of protein, carbohydrates, nucleic acids and lipids. It is classified as loosely bound and tightly bound EPS. EPS extraction techniques are involved in the segregation of the microbial cell by removing its adhesion. By extracting the EPS, the pretreatment process was enhanced by liberating the adsorbed organic biopolymers in the extracellular layer [69, 70]. EPS extraction techniques were carried out using the physical and chemical methods which was shown in Fig. 3.2

The following are the chemicals used for liberating EPS: (1) NaOH, KOH, NaCl, $CaCl_2$ and other alkaline compounds, (2) sodium citrate, citric acid and other cation binding agents and (3) TiO_2, Fenton, H_2O_2 and other AOP chemicals. A physical method such as microwave and ultrasonic waves are involved in the depletion of the EPS layer [71–76]. Disperser is also a physical method used for cell disruption for

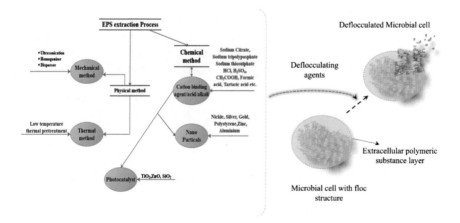

Fig. 3.2 EPS extraction techniques (Deflocculation process) and its types

enhancing the pretreatment process. By following the above methods, about 90% of EPS can be extracted which could be observed in the work of Banu et al. [77].

EPS extraction with a cation binding agent is involved in eliminating bridging ions (e.g. Ca^{2+}, Mg^{2+}, Fe^{2+} and Mg^{3+}) from the sludge flocs structure. This enables the organic release. The citric acid can extract 155 mg/L of EPS [78]. Cation binding agents are completely biodegradable in nature. Ion exchange separates monovalent (sodium) and divalent (particularly magnesium and calcium) cations from the EPS matrix, lowering floc stability and promoting floc disintegration. Using sodium citrate, nearly 93–96% of the EPS is extracted [79]. The free nitric acid (FNA) approach of EPS extraction is a sustainable and minimal chemical process that also helps to improve the digestibility of sludge. It also stimulates the synthesis of short-chain fatty acids. The deflocculating agents such as citric acid, $MgSO_4$, SDS and $MgCl_2$ have been applied at minimal concentrations acting as enzyme activators [80]. At greater doses, it acts as an enzyme inhibitor. The various intracellular elements were released into an aqueous phase by an appropriate dose of NaCl—0.03 g/g SS, resulting in a maximum sludge liquefaction rate of 23% [81–83]. Similarly, disruption of floc using low-temperature thermochemical treatment reduced the value of pH from 8 to 6.8, fostering an appropriate environment for cell lysis by Antibiotic secreting bacteria (ASB) and attaining an effective sludge solubility and dissolution rate. The cell lysis was induced by Antibiotic secreting bacteria (ASB) because it rips away the layer of peptidoglycan and proteins contained within the cell membrane of the sludge which enables the discharge of the intercellular components into the aqueous phase.

EPS extraction using AOP chemicals is a newly developed technique implemented in the field of sludge treatment. This chemical generates hydroxyl radical and causes damage to EPS [84–86]. Sharmila et al., [38] used photocatalysts to disrupt EPS. About 0.03 g/g SS of TiO_2 emits intracellular constituents into the aqueous medium with low cell lysis up to a 15 min irradiation duration increasing the liquefied level of the subsequent biological pretreatment. Fe(II)-activated persulfate ($Fe(II)-S_2O_8^{2-}$) oxidation is considered to be a novel pretreatment method that was initially introduced to improve the dewatering of waste sludge [87–89]. Persulfate ($S_2O_8^{2-}$) has been triggered to form sulphate free radicals (SO_4), which appear to be highly powerful oxidants, using heat, UV radiation, or transition metals (Me^{n+}) (redox potential 2.60 V). In ideal circumstances of about at the optimized conditions (1.5 mmol Fe(II)/g VSS, 1.2 mmol $S_2O_8^{2-}$/g VSS, and pH 3.0–8.5) it reduced the capillary suction time (CST) of the sludge up to 88.8%.

This strong oxidation promotes a better impact of SO_4 on EPS and bacterial cells. The SO_4 disrupted the reactive groups of fluorescent substances (i.e. aromatic protein-, tryptophan protein-, humic-, and fulvic-like substances) in EPS and prompted dissociation of synergies in the polymeric membrane and also instantaneous ruination of microbial species, leading to the release of EPS-bound water, intracellular ingredients, liquid hydration within the cells and possible increase of dewaterability. Sludge solubilization efficiency is improved at a specific energy input (SEI) of 171.9 kJ/kg TS when chemo-mediated ultrasonic pretreatment is used. A cation binding agent, such as citric acid, breaks the EPS in this scenario, combining both divalent cations Ca^{2+}

and Mg^{2+} and freeing intracellular components into the liquid media. The sonication pretreatment optimizes bacterial cell direct exposure during sludge solubilization as a result hydro-mechanical shear stresses were created by ultrasonic irradiation at an optimal sonic pretreatment time of 1 min [90–96].

3.3 Inhibitors Removal (Lignin)

Because of the network composition and the existence of lignin, some substrates, such as lignocellulose, are particularly resistant to AD. Numerous pretreatment procedures, including certain chemical (acid, alkali, wet oxidation), physical (separation, grinding, hydro thermolysis) and biological (microbes) methods were extensively documented in AD to boost methane output. Grinding is required in AD and composting to reduce substrate particle size. It increases the surface area, and makes the substrate more approachable to bacteria. In industrialized countries, source separation of wastes is frequently recommended prior to size reduction. Increased grinding, on the other hand, raises expenses which also potentially lowers the degrading efficiency [92]. The methane generation during AD from sisal fibre waste increased when size is reduced from 100 to 2 mm. However, in the case of sunflower oil cake, when the particle size of oil cake was reduced from 2 mm to 0.355 mm, the efficiency of VFA production decreased. The industry typically uses a greater particle size to decrease the energy consumption involved with size reduction.

Hydrothermal, chemical and biological pretreatment procedures have been developed to remove hemicellulose. Each process has advantages and disadvantages. Both hydro thermolysis and acid pretreatment remove hemicellulose and increase the accessible surface area of cellulose to microorganisms; nevertheless, the production of inhibitors such as furfural and hydroxymethylfurfural (>5 mM) is difficult. Lignin removal or modification can be attained via alkaline, moist oxidation, or biological techniques. However, the usage of alkalis (e.g. sodium) may hinder AD processes, particularly methanogenesis. Since wet oxidation is a non-selective mechanism that degrades lignin, cellulose and hemicellulose all at once, it could lead to the destruction of biodegradable components. Although certain fungi (for example, white-rot fungus) may specifically breakdown lignin under moderate circumstances, a very long time (>30 days) [97]. Because of these faults, they are unsuitable for full-scale implementation in current AD systems. The impact of hydro thermolysis and chemical pretreatment on composting is to reduce the composting time [98]. The time required to compost wood chips is reduced from 63 to 21 days after pretreatment with hydro thermolysis at 180 °C. A greater alkaline pretreatment in AD considerably reduced the rice straw disintegration time to less than 10 days from 45 days. Nevertheless, owing to the chemical and energy inputs, these pretreatments are still not suitable for composting. Chemical residues may also be a concern for digestate/compost land application.

References

1. Zhao Y, Zhao Y, Zhang Z, Wei Y, Wang H, Lu Q, Li Y, Wei Z (2017) Effect of thermo-tolerant actinomycetes inoculation on cellulose degradation and the formation of humic substances during composting. Waste Manag 68:64–73

2. Sharmila VG, Kumar MD, Pugazhendi A, Bajhaiya AK, Gugulothu P, Banu JR (2021) Biofuel production from Macroalgae: present scenario and future scope. Bioengineered 12:9216–9238

3. André L, Pauss A, Ribeiro T (2018) Solid anaerobic digestion: state-of-art, scientific and technological hurdles. Bioresour Technol 247:1027–1037

4. Monlau F, Sambusiti C, Ficara E, Aboulkas A, Barakat A, Carrère H (2015) New opportunities for agricultural digestate valorization: current situation and perspectives. Energy Environ Sci 8:2600–2621

5. Banu JR, Sharmila VG, Devi MG, Kumar SA, Kumar G, Nguyen DD, Saratale GD (2019) Cost effective sludge reduction using synergetic effect of dark fenton and disperser treatment. J Clean Prod 207:261–270

6. Degueurce A, Trémier A, Peu P (2016) Dynamic effect of leachate recirculation on batch mode solid state anaerobic digestion: Influence of recirculated volume, leachate to substrate ratio and recirculation periodicity. Bioresour Technol 216:553–561

7. Banu JR, Sharmila VG, Kavitha S, Rajajothi R, Gunasekaran M, Angappane S, Kumar G (2020) TiO2—chitosan thin film induced solar photocatalytic deflocculation of sludge for profitable bacterial pretreatment and biofuel production. Fuel 273:117741

8. Di Maria F, Barratta M, Bianconi F, Placidi P, Passeri D (2017) Solid anaerobic digestion batch with liquid digestate recirculation and wet anaerobic digestion of organic waste: comparison of system performances and identification of microbial guilds. Waste Manag 59:172–180

9. Lin L, Li Y (2017) Sequential batch thermophilic solid-state anaerobic digestion of lignocellulosic biomass via recirculating digestate as inoculum—Part I: reactor performance. Bioresour Technol 236:186–193

10. Jiang J, Liu X, Huang Y, Huang H (2015) Inoculation with nitrogen turnover bacterial agent appropriately increasing nitrogen and promoting maturity in pig manure composting. Waste Manag 39:78–85

11. Sharmila VG, Kavitha S, Rajashankar K, Yeom IT, Banu JR (2015) Effects of titanium dioxide mediated dairy waste activated sludge deflocculation on the efficiency of bacterial disintegration and cost of sludge management. Bioresour Technol 197:64–71

12. Zeng G, Yu M, Chen Y, Huang D, Zhang J, Huang H, Jiang R, Yu Z (2010) Effects of inoculation with *Phanerochaete chrysosporium* at various time points on enzyme activities during agricultural waste composting. Bioresour Technol 101:222–227

13. Xi B, He X, Dang Q, Yang T, Li M, Wang X, Li D, Tang J (2015) Effect of multi-stage inoculation on the bacterial and fungal community structure during organic municipal solid wastes composting. Bioresour Technol 196:399–405

14. Sung S, Liu T (2003) Ammonia inhibition on thermophilic anaerobic digestion. Chemosphere 53:43–52

15. Shi J, Wang Z, Stiverson JA, Yu Z, Li Y (2013) Reactor performance and microbial community dynamics during solid-state anaerobic digestion of corn stover at mesophilic and thermophilic conditions. Bioresour Technol 136:574–581

16. Keener HM, Ekinci K, Michel FC (2005) Composting process optimization—using on/off controls. Compost Sci Util 13:288–299

17. Li Q, Wang XC, Zhang HH, Shi HL, Hu T, Ngo HH (2013) Characteristics of nitrogen transformation and microbial community in an aerobic composting reactor under two typical temperatures. Bioresour Technol 137:270–277

18. Liang C, Das KC, McClendon RW (2003) The influence of temperature and moisture contents regimes on the aerobic microbial activity of a biosolids composting blend. Bioresour Technol 86:131–137

19. Brown D, Shi J, Li Y (2012) Comparison of solid-state to liquid anaerobic digestion of lignocellulosic feedstocks for biogas production. Bioresour Technol 124:379–386

20. Yang L, Li Y (2014) Anaerobic digestion of giant reed for methane production. Bioresour Technol 171:233–239

21. Das KC (2008) Co-composting of alkaline tissue digester effluent with yard trimmings. Waste Manag 28:1785–1790

22. Bhatia SK, Jagtap SS, Bedekar AA, Bhatia RK, Patel AK, Pant D, Banu JR, Rao CV, Kim Y-G, Yang Y-H (2020) Recent developments in pretreatment technologies on lignocellulosic biomass: effect of key parameters, technological improvements, and challenges. Bioresour Technol 300:122724

23. Banu JR, Sokkanathan G, Sharmila VG, Tamilarasan K, Kumar SA, Jamal MT (2018) Combinative treatment of chocolaterie wastewater by a hybrid up-flow anaerobic sludge blanket reactor and solar photo Fenton process. Desalin Water Treat 121:343–350

24. Madigan MT, Martinko JM (2006) Brock biology of microorganisms, 11th edn. Prentice Hall, Pearson

25. Banu JR, Sharmila VG, Ushani U, Amudha V, Kumar G (2020) Impervious and influence in the liquid fuel production from municipal plastic waste through thermo-chemical biomass conversion technologies: a review. Sci Total Environ 718:137287

26. Torres-Climent A, Martin-Mata J, Marhuenda-Egea F, Moral R, Barber X, Perez-Murcia MD, Paredes C (2015) Composting of the solid phase of digestate from biogas production: optimization of the moisture, C/N ratio, and pH conditions. Commun Soil Sci Plant Anal 46:197–207

27. Xu F, Shi J, Lv W, Yu Z, Li Y (2013) Comparison of different liquid anaerobic digestion effluents as inocula and nitrogen sources for solid-state batch anaerobic digestion of corn stover. Waste Manag 33:26–32

28. Sharmila VG, Gunasekaran M, Angappane S, Zhen G, Tae Yeom I, Banu JR (2019) Evaluation of photocatalytic thin film pretreatment on anaerobic degradability of exopolymer extracted biosolids for biofuel generation. Bioresour Technol 279:132–139

29. Wang K, Yin J, Shen D, Li N (2014) Anaerobic digestion of food waste for volatile fatty acids (VFAs) production with different types of inoculum: effect of pH. Bioresour Technol 161:395–401

30. Demirel B, Scherer P (2008) The roles of acetotrophic and hydrogenotrophic methanogens during anaerobic conversion of biomass to methane: a review. Rev Environ Sci Biotechnol 7:173–190

31. Xu S, Selvam A, Wong JWC (2014) Optimization of micro-aeration intensity in acidogenic reactor of a two-phase anaerobic digester treating food waste. Waste Manag 34:363–369

32. Sharmila VG, Banu JR, Gunasekaran M, Angappane S, Yeom IT (2018) Nano-layered TiO_2 for effective bacterial disintegration of waste activated sludge and biogas production: immobilized TiO_2 mediated bacterial pretreatment of WAS for anaerobic digestion. J Chem Technol Biotechnol 93:2701–2709

33. Fu S-F, Wang F, Shi X-S, Guo R-B (2016) Impacts of microaeration on the anaerobic digestion of corn straw and the microbial community structure. Chem Eng J 287:523–528

34. Nghiem LD, Manassa P, Dawson M, Fitzgerald SK (2014) Oxidation reduction potential as a parameter to regulate micro-oxygen injection into anaerobic digester for reducing hydrogen sulphide concentration in biogas. Bioresour Technol 173:443–447

35. Zhu M, Lü F, Hao L-P, He P-J, Shao L-M (2009) Regulating the hydrolysis of organic wastes by micro-aeration and effluent recirculation. Waste Manag 29:2042–2050

36. Krayzelova L, Bartacek J, Díaz I, Jeison D, Volcke EIP, Jenicek P (2015) Microaeration for hydrogen sulfide removal during anaerobic treatment: a review. Rev Environ Sci Biotechnol 14:703–725

37. Ahn HK, Richard TL, Choi HL (2007) Mass and thermal balance during composting of a poultry manure wood shavings mixture at different aeration rates. Process Biochem 42:215–223

38. Yeom IT, Sharmila VG, Kannah RY, Sivashanmugam P, Banu JR (2018) Municipal waste management. In: Bryant B, Hall B (eds) Municipal and industrial waste: source, management practices and future challenges. Nova Science Publishers, Inc, pp 181–224

39. Doublet J, Francou C, Poitrenaud M, Houot S (2011) Influence of bulking agents on organic matter evolution during sewage sludge composting; consequences on compost organic matter stability and N availability. Bioresour Technol 102:1298–1307
40. Chang JI, Chen YJ (2010) Effects of bulking agents on food waste composting. Bioresour Technol 101:5917–5924
41. Preethi BJR, Sharmila VG, Kavitha S, Varjani S, Kumar G, Gunasekaran M (2021) Alkali activated persulfate mediated extracellular organic release on enzyme secreting bacterial pretreatment for efficient hydrogen production. Bioresour Technol 341:125810
42. Eswari AP, Sharmila VG, Gunasekaran M, Banu JR (2020) New business and marketing concepts for cross-sector valorization of food waste. In: Food waste to valuable resources. Elsevier, pp 417–433
43. Puyuelo B, Gea T, Sánchez A (2010) A new control strategy for the composting process based on the oxygen uptake rate. Chem Eng J 165:161–169
44. Mejias L, Komilis D, Gea T, Sánchez A (2017) The effect of airflow rates and aeration mode on the respiration activity of four organic wastes: Implications on the composting process. Waste Manag 65:22–28
45. Wei L, Shutao W, Jin Z, Tong X (2014) Biochar influences the microbial community structure during tomato stalk composting with chicken manure. Bioresour Technol 154:148–154
46. Sánchez-García M, Alburquerque JA, Sánchez-Monedero MA, Roig A, Cayuela ML (2015) Biochar accelerates organic matter degradation and enhances N mineralisation during composting of poultry manure without a relevant impact on gas emissions. Bioresour Technol 192:272–279
47. Ahmed B, Tyagi S, Banu R, Kazmi AA, Tyagi VK (2021) Carbon based conductive materials mediated recalcitrant toxicity mitigation during anaerobic digestion of thermo-chemically pretreated organic fraction of municipal solid waste. Chemosphere 132682
48. Bhatia SK, Otari SV, Jeon J-M, Gurav R, Choi Y-K, Bhatia RK, Pugazhendhi A, Kumar V, Banu JR, Yoon J-J, Choi K-Y, Yang Y-H (2021) Biowaste-to-bioplastic (polyhydroxyalkanoates): conversion technologies, strategies, challenges, and perspective. Bioresour Technol 326:124733
49. Bruno LB, Anbuganesan V, Karthik C, Tripti KA, Banu JR, Freitas H, Rajkumar M (2021) Enhanced phytoextraction of multi-metal contaminated soils under increased atmospheric temperature by bioaugmentation with plant growth promoting *Bacillus cereus*. J Environ Manage 289:112553
50. Ginni K, Kannah RY, Bhatia SK, Kumar A, Rajkumar KG, Pugazhendhi A, Chi NTL, Banu R (2021) Valorization of agricultural residues: different biorefinery routes. J Environ Chem Eng 9:105435
51. Gopikumar S, Banu JR, Robinson YH, Shanmuganathan V, Kadry S, Rho S (2021) Novel framework of GIS based automated monitoring process on environmental biodegradability and risk analysis using Internet of Things. Environ Res 194:110621
52. Khan MJ, Singh N, Mishra S, Ahirwar A, Bast F, Varjani S, Schoefs B, Marchand J, Rajendran K, Banu JR, Saratale GD, Saratale RG, Vinayak V (2022) Impact of light on microalgal photosynthetic microbial fuel cells and removal of pollutants by nanoadsorbent biopolymers: updates, challenges and innovations. Chemosphere 288:132589
53. Pugazhendi A, Alreeshi GG, Jamal MT, Karuppiah T, Jeyakumar RB (2021) Bioenergy production and treatment of aquaculture wastewater using saline anode microbial fuel cell under saline condition. Environ technol innov 21:101331
54. Raj T, Chandrasekhar K, Banu R, Yoon J-J, Kumar G, Kim S-H (2021) Synthesis of γ-valerolactone (GVL) and their applications for lignocellulosic deconstruction for sustainable green biorefineries. Fuel 303:121333
55. Sethupathy A, Kumar PS, Sivashanmugam P, Arun C, Banu JR, Ashokkumar M (2021) Evaluation of biohydrogen production potential of fragmented sugar industry biosludge using ultrasonication coupled with egtazic acid. Int J Hydrogen Energy 46:1705–1714
56. Sethupathy A, Pathak PK, Sivashanmugam P, Arun C, Banu JR, Ashokkumar M (2021) Enrichment of hydrogen production from fruit waste biomass using ozonation assisted with citric acid. Waste Manag Res 734242X211010364

57. Sim Y-B, Jung J-H, Baik J-H, Park J-H, Kumar G, Banu JR, Kim S-H (2021) Dynamic membrane bioreactor for high rate continuous biohydrogen production from algal biomass. Bioresour Technol 340:125562

58. Tyagi VK, Kapoor A, Arora P, Banu JR, Das S, Pipesh S, Kazmi AA (2021) Mechanical-biological treatment of municipal solid waste: case study of 100 TPD Goa plant, India. J Environ Manage 292:112741

59. Yap JK, Sankaran R, Chew KW, Halimatul Munawaroh HS, Ho S-H, Banu JR, Show PL (2021) Advancement of green technologies: a comprehensive review on the potential application of microalgae biomass. Chemosphere 281:130886

60. Jamal MT, Pugazhendi A, Banu JR (2020) Application of halophiles in air cathode MFC for seafood industrial wastewater treatment and energy production under high saline condition. Environ technol innov 20:101119

61. Pugazhendi A, Al-Mutairi AE, Jamal MT, Banu JR, Palanisamy K (2020) Treatment of seafood industrial wastewater coupled with electricity production using air cathode microbial fuel cell under saline condition. Int J Energy Res 44:12535–12545

62. Banu JR, Tamilarasan K, Chang SW, Nguyen DD, Ponnusamy VK, Kumar G (2020) Surfactant assisted microwave disintegration of green marine macroalgae for enhanced anaerobic biodegradability and biomethane recovery. Fuel (Lond) 281:118802

63. Rani G, Nabi Z, Banu JR, Yogalakshmi KN (2020) Batch fed single chambered microbial electrolysis cell for the treatment of landfill leachate. Renew Energy 153:168–174

64. Selvaraj D, Somanathan A, Banu JR, Kumar G (2020) Generation of electricity by the degradation of electro-Fenton pretreated latex wastewater using double chamber microbial fuel cell. Int J Energy Res 44:12496–12505

65. Sethupathy A, Arunagiri A, Sivashanmugam P, Banu JR, Ashokkumar M (2020) Disperser coupled rhamnolipid disintegration of pulp and paper mill waste biosolid: characterisation, methane production, energy assessment and cost analysis. Bioresour Technol 297:122545

66. Shahid MK, Kashif A, Rout PR, Aslam M, Fuwad A, Choi Y, Banu JR, Park JH, Kumar G (2020) A brief review of anaerobic membrane bioreactors emphasizing recent advancements, fouling issues and future perspectives. J Environ Manage 270:110909

67. Banu JR, Kumar MD, Gunasekaran M, Kumar G (2019) Biopolymer production in bio electrochemical system: literature survey. Bioresour technol rep 7:100283

68. Dinesh MD, Kaliappan S, Gopikumar S, Zhen G, Banu JR (2019) Synergetic pretreatment of algal biomass through H_2O_2 induced microwave in acidic condition for biohydrogen production. Fuel (Lond) 253:833–839

69. Harinee S, Muthukumar K, Dahms H-U, Koperuncholan M, Vignesh S, Banu JR, Ashok M, James RA (2019) Biocompatible nanoparticles with enhanced photocatalytic and anti-microfouling potential. Int Biodeterior Biodegradation 145:104790

70. Jeong SY, Chang SW, Ngo HH, Guo W, Nghiem LD, Banu JR, Jeon B-H, Nguyen DD (2019) Influence of thermal hydrolysis pretreatment on physicochemical properties and anaerobic biodegradability of waste activated sludge with different solids content. Waste Manag 85:214–221

71. Kumar G, Ponnusamy VK, Bhosale RR, Shobana S, Yoon J-J, Bhatia SK, Banu JR, Kim S-H (2019) A review on the conversion of volatile fatty acids to polyhydroxyalkanoates using dark fermentative effluents from hydrogen production. Bioresour Technol 287:121427

72. Nguyen DD, Jeon B-H, Jeung JH, Rene ER, Banu JR, Ravindran B, Vu CM, Ngo HH, Guo W, Chang SW (2019) Thermophilic anaerobic digestion of model organic wastes: evaluation of biomethane production and multiple kinetic models analysis. Bioresour Technol 280:269–276

73. Nguyen XC, Chang SW, Tran TCP, Nguyen TTN, Hoang TQ, Banu JR, Al-Muhtaseb AH, La DD, Guo W, Ngo HH, Nguyen DD (2019) Comparative study about the performance of three types of modified natural treatment systems for rice noodle wastewater. Bioresour Technol 282:163–170

74. Pan Y, Zhi Z, Zhen G, Lu X, Bakonyi P, Li Y-Y, Zhao Y, Banu JR (2019) Synergistic effect and biodegradation kinetics of sewage sludge and food waste mesophilic anaerobic co-digestion and the underlying stimulation mechanisms. Fuel 253:40–49

75. Ponnusamy VK, Nguyen DD, Dharmaraja J, Shobana S, Banu JR, Saratale RG, Chang SW, Kumar G (2019) A review on lignin structure, pretreatments, fermentation reactions and biorefinery potential. Bioresour Technol 271:462–472

76. Pugazhendhi A, Shobana S, Bakonyi P, Nemestóthy N, Xia A, Banu JR, Kumar G (2019) A review on chemical mechanism of microalgae flocculation via polymers. Biotechnol Rep 21:e00302

77. Banu JR, Tamilarasan K, Rani RU, Gunasekaran M, Cho S-K, Al-Muhtaseb AH, Kumar G (2019) Dispersion aided tenside disintegration of seagrass Syringodium isoetifolium: towards biomethanation, kinetics, energy exploration and evaluation. Bioresour Technol 277:62–67

78. Shanthi M, Banu JR, Sivashanmugam P (2019) Synergistic effect of combined pretreatment in solubilizing fruits and vegetable residue for biogas production: hydrolysis, energy assessment. Fuel 250:194–202

79. Subha C, Kavitha S, Abisheka S, Tamilarasan K, Arulazhagan P, Banu JR (2019) Bioelectricity generation and effect studies from organic rich chocolaterie wastewater using continuous upflow anaerobic microbial fuel cell. Fuel 251:224–232

80. Antonopoulou G, Alexandropoulou M, Lytras C, Lyberatos G (2015) Modeling of anaerobic digestion of food industry wastes in different bioreactor types. Waste Biomass Valorization 6:335–341

81. Appels L, Baeyens J, Degrève J, Dewil R (2008) Principles and potential of the anaerobic digestion of waste-activated sludge. Prog Energy Combust Sci 34:755–781

82. Arumugam T, Parthiban L, Rangasamy P (2015) Two-phase anaerobic digestion model of a tannery solid waste: experimental investigation and modeling with ANFIS. Arab J Sci Eng 40:279–288

83. Batstone DJ (2006) Mathematical modelling of anaerobic reactors treating domestic wastewater: rational criteria for model use. Rev Environ Sci Biotechnol 5:57–71

84. Batstone DJ, Puyol D, Flores-Alsina X, Rodríguez J (2015) Mathematical modelling of anaerobic digestion processes: applications and future needs. Rev Environ Sci Biotechnol 14:595–613

85. Behera SK, Meher SK, Park H-S (2015) Artificial neural network model for predicting methane percentage in biogas recovered from a landfill upon injection of liquid organic waste. Clean Technol Environ Policy 17:443–453

86. Bernard O, Hadj-Sadok Z, Dochain D, Genovesi A, Steyer JP (2001) Dynamical model development and parameter identification for an anaerobic wastewater treatment process. Biotechnol Bioeng 75:424–438

87. Blumensaat F, Keller J (2005) Modelling of two-stage anaerobic digestion using the IWA Anaerobic Digestion Model No. 1 (ADM1). Water Res 39:171–183

88. Bravo AD, Mailier J, Martin C, Rodríguez J, Lara CAA, Wouwer AV (2011) Model selection, identification and validation in anaerobic digestion: a review. Water Res 45:5347–5364

89. Bryers JD (1985) Structured modeling of the anaerobic digestion of biomass particulates. Biotechnol Bioeng 27:638–649

90. Donoso-Bravo A, Pérez-Elvira SI, Fdz-Polanco F (2010) Application of simplified models for anaerobic biodegradability tests. Evaluation of pre-treatment processes. Chem Eng J 160:607–614

91. Ebenezer AV, Arulazhagan P, Kumar SA, Yeom IT, Banu JR (2015) Effect of deflocculation on the efficiency of low-energy microwave pretreatment and anaerobic biodegradation of waste activated sludge. Appl Energy 145:104–110

92. Raj SE, Banu JR, Kaliappan S, Yeom IT, Kumar SA (2013) Effects of side-stream, low temperature phosphorus recovery on the performance of anaerobic/anoxic/oxic systems integrated with sludge pretreatment. Bioresour Technol 140:376–384

93. Fedorovich V, Lens P, Kalyuzhnyi S (2003) Extension of anaerobic digestion model no. 1 with processes of sulfate reduction. Appl Biochem Biotechnol 109:33–45

94. Feng Y, Behrendt J, Wendland C, Otterpohl R (2006) Parameter analysis of the IWA anaerobic digestion model no. 1 for the anaerobic digestion of blackwater with kitchen refuse. Water Sci Technol 54:139–214

95. Ganidi N, Tyrrel S, Cartmell E (2009) Anaerobic digestion foaming causes–a review. Bioresour Technol 100:5546–5554
96. Gavala HN, Angelidaki I, Ahring BK (2003) Kinetics and modeling of anaerobic digestion process. Adv Biochem Eng Biotechnol 81:57–93
97. Grant WD, Lawrence TM (2014) A simplified method for the design and sizing of anaerobic digestion systems for smaller farms. Environ Dev Sustain 16:345–360
98. Hill DT (1982) A comprehensive dynamic model for animal waste methanogenesis. Trans ASAE 25:1374–1380

Chapter 4
Hydrolysis and Assessment

4.1 Introduction

The pretreatment aids hydrolysis and is considered as a challenging process in recovering energy from biomass. During pretreatment, modification occurs at the macroscopic and microscopic levels of biomass such as size, structure, and chemical configuration. Pretreatment aids the hydrolysis of biomass and enhances the efficiency of the subsequent biological and chemical conversion process. Figure 4.1 shows the impact of biological pretreatment of sludge on fermentation. From the figure, it was evident that as a result of pretreatment, fermentation of substrate is enhanced. This resulted in the production of more volatile fatty acids. Biomass pretreatment is an essential step and it enhances the energy extracting potential of biological and chemical conversion processes. It has a greater influence on the biodegradation of biomass, cost of downstream which involves removal of toxic substances, enzyme loading, the demand for waste treatment, and other parameters. Typically, some forms of pretreatment are physical (mechanical), chemical, biological, and combinative (thermochemical or biochemical) methods.

For an effective and absolute hydrolysis, biomass has to be pretreated prior to biological and chemical conversion processes. The pretreatment methods remove the recalcitrant and unwanted substance from the biomass and improve the energy conversion efficiency of biological and chemical processes. The pretreatment of biomass is considered as a first step and the most challenging process in recovering energy from biomass. This pre-treatment is an essential step to enhance the energy extracting potential of biological and chemical conversion processes. It has a greater influence on the biodegradation of biomass, cost of downstream which involves removal of toxic substances, enzyme loading, the demand for waste treatment, and other parameters. Typically, some forms of pretreatment are physical (mechanical), chemical, biological, and combinative (thermochemical or biochemical) methods.

© Springer Nature Singapore Pte Ltd. 2022
K. Sudalyandi and R. Jeyakumar, *Biofuel Production Using Anaerobic Digestion*,
Green Energy and Technology, https://doi.org/10.1007/978-981-19-3743-9_4

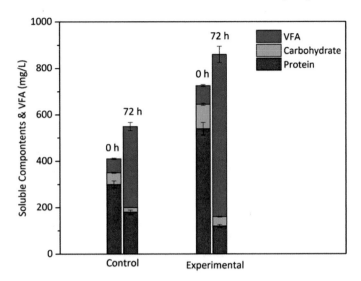

Fig. 4.1 Impact of pretreatment on sludge fermentation

4.2 Physical Pretreatment

A physical pretreatment method comprises mainly mechanical processes such as chipping, grinding, and milling. Figure 4.2 shows the schematic representation of various size reduction method. The mechanical process which includes reduction of size by shredding and grinding is the first step in the pretreatment of biomass such as wood waste and straw. Mechanical pretreatments permit the split-up of the chief complex plant portions into simple portions such as tissues, cells, and polymers, which are utilized as biomass for numerous applications. By mechanical pretreatment, the size of biomass can be reduced and is an essential process for converting biomass waste into energy [1]. There are various size reduction methods that are differentiated according to the size of the particle.

For reducing the particle size in the range of millimetre to centimetre, the biomass can be subjected to the cutting process. For reducing the size from 15 cm to 500 mm, the biomass can be subjected to coarse milling [2]. Reduction of size from cm to 100 mm can be achieved through intermediate micronization. The size of biomass can be reduced to less than 100, 30, and 1 mm by fine, ultrafine, and nano grinding.

In mechanical pretreatment, size reduction is attained through combined mechanistic actions which include impact, solidity, abrasion, and shear. It has numerous benefits:

- It improves the calorific value of waste biomass.
- It simplifies the raw material supply chain and its storage circumstances.
- It increases the availability of biomass for hydrolysis by decreasing the overall accessible surface area.

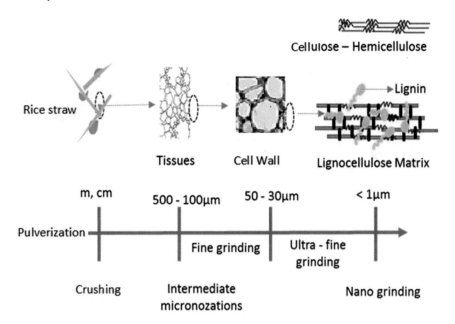

Fig. 4.2 Mechanical size reduction process

- It decreases the mass and heat transfer and consequently reduces energy inputs.
- Transportation of biomass from the place of production to the place of utilization occurs more efficiently.

The selection of equipment is mainly influenced by various parameters such as physiochemical characteristics of biomass (moisture and size of the particle) and appliance target. The choice of mechanical equipment is considered as the most essential economical parameter due to its energy necessity in relative to final particle size.

It chiefly depends on the following:

- Instrument specifications which include the speed of the motor.
- Input material characteristics.
- Physiochemical properties and biomass structure.
- Particle sizes.

Hardwoods need more energy input than agricultural residues. Due to particle size reduction by mechanical pretreatment, it is utilized as a pre-pretreatment step for all other pretreatment methods such as chemical and biological pretreatments. High-pressure homogenizer is one of the maximum used mechanical treatments for large-scale setup. Figure 4.3 shows the mechanism of high-pressure homogenizer on algae biomass disintegration. High-pressure homogenization is a purely mechanical process, which is induced by imposing a fluidic product through a thin gap (the homogenizing nozzle) at high pressure (150–200 MPa).

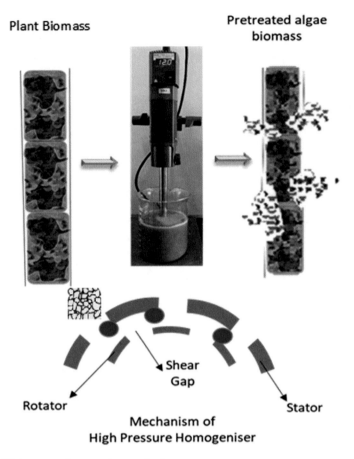

Fig. 4.3 High-pressure homogenizer mechanism

The biomass is exposed to these narrow gaps to obtain cavitation and shear pressures, resulting in cell disintegration. During high-pressure homogenization, the pressure increases to a maximal level (bar) and homogenizes the biomass through high depressurization. The produced cavitation provokes mechanical energy which breaks the cell wall. The extent of homogenization for pretreating the biomass is governed by its specific energy consumption.

Arriving optimum specific energy is essential for cost-effective biofuel production. The specific energy utilization of homogenization is calculated by the following equation:

$$SE = P * t/V * TS \tag{4.1}$$

where P is the power consumed during homogenization (kW), t is the time of homogenization (s), V is the volume of biomass (L), and TS is the total solids of the biomass (kg/L).

The rpm is a crucial influencing element during high-pressure homogenization (HPH). To increase efficiency, this HPH approach might be combined with others.

As a result of liquefaction, intracellular release of organics occurs. The liquefaction was calculated by the following equation:

$$\alpha \frac{(SCODp - SCODi)}{(TCODi - SCODi)} \tag{4.2}$$

where

α = solubilization efficiency (%);
SCODp = Concentration of soluble COD after pretreatment (mg/L);
SCODi = Concentration of soluble COD prior to pretreatment (mg/L);
TCODi = Concentration of total COD prior to pretreatment (mg/L).

The mechanical pretreatment enhances the hydrolysis of biomass through liquefaction. The various physical or mechanical pretreatment methods include milling, ultrasound, and microwave pretreatments. Table 4.1 lists the usage of high-pressure homogenization in pretreating biomass.

Table 4.1   Usage of high-pressure homogenization in pretreating biomass

S. No.	Biomass used	Pretreatment condition	Conversion process	Energy produced	References
1	Municipal WAS	600 bars	Anaerobic digestion	Biogas	[3]
2	Municipal WAS	12,000 psi	Anaerobic digestion	Methane	[4]
3	Municipal WAS	ΔP > 30 bar	Anaerobic digestion	Biogas	[1]
4	Food Waste	120 °C, 10 bar for 30 min	Anaerobic digestion	Methane	[2]
5	Pulp mill WAS	4000 kJ/kg TS	Anaerobic digestion	Methane	[5]
6	Dairy WAS	12,000 rpm; 4544 kJ/kg TS	Anaerobic digestion	Methane	[6]
7	Dairy WAS	12,000 rpm; 5013 kJ/kg TS	Anaerobic digestion	Biogas	[7]
8	Plant Biomass	10 MPa	Fermentation	Bioethanol	[8]
9	Microalgae	206.84 MPa	Fermentation	Bioethanol	[9]
10	Dairy WAS	12,000 rpm; 7377 kJ/kg TS	Anaerobic digestion	Methane	[10]

rpm—Revolution per minute; WAS—Waste activated sludge; SE—Specific energy; TS—Total solids

4.2.1   Bead Milling

Bead milling mechanically cracks the cell wall of biomass through agitation and grinding on a solid surface with glass beads. Incitation of beads produces mechanical shear forces that disrupt the cell walls of biomass. Figure 4.4 shows the shaking vessel bead milling pretreatment. The optimum diameter of the bead size should be between 0.3 and 0.5 mm for efficient biomass disintegration. The beads used for biomass disintegration are made up of zirconia silica, zirconium oxide, or titanium carbide.

The factors that depend upon the bead milling process are the concentration of the biomass, shape, size of beads, and contact between biomass and the beads. There are two types of bead milling namely

- Shaking vessel bead milling and
- Agitated bead milling.

Shaking vessel bead milling

In this type of bead milling, the entire shaking vessels are shaken to disrupt the cell walls of biomass. During this pretreatment, single or multiple vessels that are placed on a stage vibrate forcing the beads to move towards the biomass. A collision occurs between the beads and the biomass. This method is used for lipid extraction from microalgal biomass for biodiesel production. This method is not effective in destroying the cell walls of biomass. So its application is limited to the laboratory level.

Agitated bead milling

The agitated bead mill is more suitable for disruption and ultrafine grinding of biomass. During this pretreatment, both the beads and biomass are subjected to

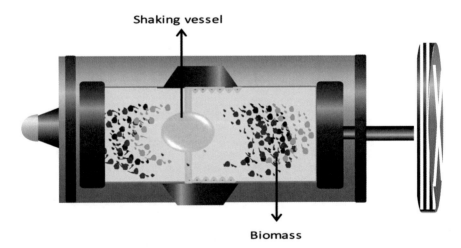

Fig. 4.4 Shaking vessel milling process

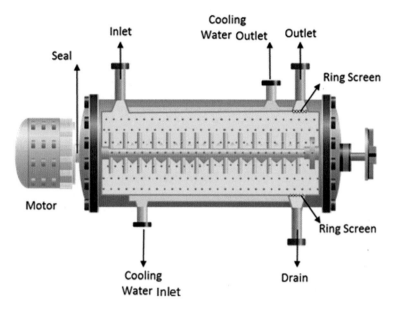

Fig. 4.5 Agitated bead milling process

agitation. Heat is provided by a rotating agitator inside the vessel. Agitating speed is the main parameter that increases the shear force and frequency to a greater extent. This in turn enhances the breakage of biomass. These bead mills have a dynamic separating system, so that the grinded materials can be easily separated with no clogging. Figure 4.5 shows the agitated bead milling process. The vessel is composed of a cooling jacket for pretreating heat-susceptible biomass. These types of bead mills are best suited to extract bio-oil and lipids for biodiesel production.

4.2.2 Ultrasound Pretreatment

Ultrasonic pretreatment is a mechanical pretreatment in which the disintegration is accomplished through the contact of high-frequency sound waves produced by a vibrating horn. When biomass is subjected to ultrasound pretreatment, the local pressure in the liquid phase is reduced below the evaporating pressure and leads to the formation of microbubbles [10–12]. These microbubbles in turn undergo hasty collapse and expansion due to elevated thermal and pressure gradients in the aqueous phase. This leads to the breakage of the biomass cell wall and the discharge of cellular components into the aqueous medium. In this way, ultrasonic pretreatment generates mechanical shear forces by cavitation effect. Parameters such as biomass thickness, surrounding pressure, and liquefied gas quantity can affect the extent of cavitation generated through ultrasound. Sonic waves are employed at lesser (<50 kHz) and greater (>50 kHz) frequencies. Lesser frequency induces physical impacts, and greater frequency induces the generation of hydroxyl radicals. Once the frequency has been set constant, the variables that influence the ultrasonic sonic pretreatment are power and time. Specific energy is the main governing parameter of ultrasound disintegration. It is estimated to be the product of power consumption and pretreatment time, divided by the amount of biomass (i.e. solid concentration) and volume. The specific energy is calculated by the following equation:

$$\text{Specific energy (kJ /kg TS)} = P(W) \times t(s)/V(L) \times TS(g\,L^{-1})$$

where P is the power, t is the treatment time, V is the volume of the biomass waste, and T is the total solids of the biomass waste. The biomass concentration has a significant impact on the efficiency of ultrasonic disintegration. Increasing or decreasing the biomass concentration below to threshold limit has a negative effect on SE consumption. Hence, optimization of biomass concentration is a vital step for employing ultrasonic pretreatment for energetically positive biofuel production. Temperature is also considered to be a crucial parameter, as it influences the vapour pressure within the cell. As the temperature is lowered, the pressure also got decreased increasing the efficiency of ultrasound pretreatment. Figure 4.6 shows the mechanism of ultrasound pre-treatment.

Cavitation bubbles have physical and chemical effects on the reaction medium. A physical effect is produced by forceful native convection in the medium and it increases the mass transfer characteristics. Radicals generated inside the cavitational bubble get released into the medium with the fragmentation of the bubble which in turn enhances chemical reactions. The physical and chemical effects produced by ultrasound were found to improve the pretreatment of lignocellulose via delignification and surface erosion. Ultrasonication of sludge biomass enhances a 50% rise in the biogas generation. Table 4.2 lists the usage of ultrasonication in pretreating biomass. Ultrasonic disintegration generates cavitation bubbles, which aid in the disintegration of the cell membranes, resulting in enhanced intracellular discharge.

Sonication duration and frequency are critical parameters for attaining maximal sludge solubilization. The sludge solubilization varies at different sonication times. To achieve 75–80% of solubilization, 90 min sonication was necessary which requires 30–40 min ultrasonication to reach 50% solubilization. Nonetheless, this method improves cell wall breakdown and organic matter solubilization in microalgae, though the benefits vary depending on the microalgae species and pretreatment circumstances, such as the applied specific energy. Scenedesmus sp. methane output rose by 14% with a specific energy of 76.5 MJ/kg TS, whereas it increased by 75–88% with a specific energy of 100–130 MJ/kg TS [13].

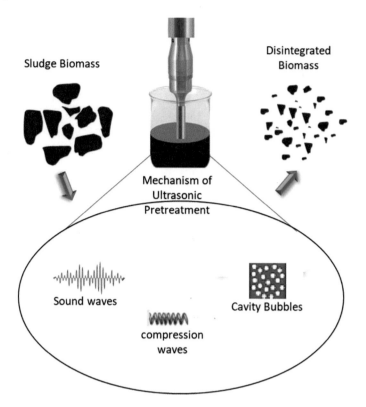

Fig. 4.6 Liquefaction through ultrasound

4.2.3 Microwave Pretreatment

Biomass pretreatment by microwave is an alternative to traditional heat pretreatment. Figure 4.7 shows the microwave pretreatment process. Microwave is an electromagnetic irradiation with a wavelength ranging from 1 mm to 1 m, equivalent to an oscillation frequency of 300 MHz to 0.3 GHz. Domestic and industrial microwaves usually function at 2.45 GHz frequency with a wavelength of about 12.24 cm and an irradiation energy of about 1.02×10^{-5} eV [18]. The mechanism of microwave-mediated pretreatment of biomass happens via thermal and athermal. The thermal effect is induced through the contact of fluctuating electrical field with bipolar compounds such as water and soluble biopolymers. The inter and intramolecular movements induce thermal energy which increases the temperature and pressure impacts in the cell wall of biomass and causes rupture.

The athermal effect induces rapid alteration in the dipole arrangement of cell wall biopolymers that causes the cleavage of hydrogen bonds. This in turn paves the way for biomass disintegration and alterations in the structural arrangement of proteins in biomass. Microwave frequency, biomass quantity, exposure time, temperature, and

Table 4.2 Usage of ultrasonication in pretreating biomass

S. No.	Biomass used	Pretreatment condition	Conversion process	Energy produced	References
1	Cassava chip	Probe; 20 kHz, 8.5 W/mL; 10–30 s	Fermentation	Bioethanol	[10]
2	Sugarcane bagasse	45 kHz, about 0.01 W/mL; 72 h	Transesterification	Biodiesel	[11]
3	Vegetable oils	Probe, 20 kHz, 3 W/mL, 45–65 °C, 2–5 h	Transesterification	Biodiesel	[13]
4	Municipal solid waste	Sonication at 20 kHz for 30–60 min	Anaerobic digestion	Methane	[12]
5	Microalgae	234 MJ/kg VS	Anaerobic digestion	Methane	[2]
6	Food waste	13,500 kJ/kg TS (SE)	Fermentation	Hydrogen	[14]
7	Chicken manure	1kWh/kg ODM (SE)	Anaerobic digestion	Biogas	[15]
8	Dairy WAS	171.9 kJ/kg TS (SE)	Anaerobic digestion	Methane	[16]
9	Municipal WAS	641 kJ/kg TS (SE)	Anaerobic digestion	Methane	[17]
10	Plant biomass	100 W ultrasonic irradiation; 45 min	Fermentation	Bioethanol	[9]

WAS—Waste activated sludge; ODM—Organic dry matter; TS—Total solids; SE—Specific energy

penetration intensity are the main parameters that influence the disintegration potential microwave pretreatment. Microwave pretreatment similar to sonication revealed that the pretreatment effect on biomass solubilization and methane increased with the applied specific energy, regardless of output power or exposure time. In continuous reactors running at 20 days HRT, methane output was 60% higher after microwave pretreatment (0.27 L CH4/g VS) compared to the control (0.17 L CH4/g VS). Similar to ultrasonication, the major driving parameters that affect microwave disintegration are applied power and exposure time [19].

Rapid penetration of irradiation power into biomass is the main benefit of this pretreatment process. The main disadvantage is elevated energy consumption. It is one of the most effective biomass pretreatment processes that modifies the structure and composition of cellulose, lignin, and hemicelluloses, which in turn enhances the efficiency of the WTE process. Table 4.3 lists the usage of microwaves in pretreating biomass.

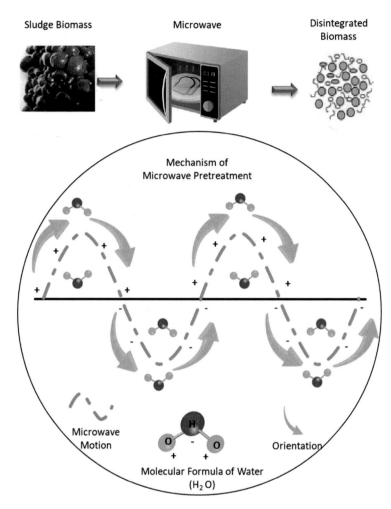

Fig. 4.7 Liquefaction through microwave

4.3 Chemical Pretreatment

Chemical pretreatment is mostly applied to plant and waste activated sludge biomass as it helps in disintegrating the cell wall. A shielding layer called lignin around the cellulose protects the biomass from hydrolytic enzyme attack. Removal of lignin by chemical pretreatment facilitates the accessibility of cellulose and hemicelluloses for the action of enzymes. This results in increased production of biofuel. Some of the most promising chemical agents used to pretreat biomass include acids, alkali, salts, oxidants, and solvents [19]. Figure 4.8 shows the mechanism of chemical pretreatment. Based on pH, chemical pretreatment was classified into acidic and alkaline pretreatments. The main objectives of chemical pretreatment are as follows:

Table 4.3 Usage of microwaves in pretreating biomass

S. No.	Biomass used	Pretreatment condition	Conversion process	Energy produced	References
1	Food waste	MW intensity 7.8 °C/min	Anaerobic digestion	Biogas	[18]
2	Microalgae	60 Wh/g of dried biomass	Transesterification	Biodiesel	[19]
3	Municipal solid waste	115–145 °C for 40 min	Anaerobic digestion	Biogas	[20]
4	Microalgae	0.9 Wh/g to 1.6 Wh/g of biomass	Transesterification	Biodiesel	[21]
5	Dairy WAS	1844 kJ/L (IE)	Anaerobic digestion	Biogas	[6]
6	Municipal WAS	96 kJ/kg TS (SE)	Anaerobic digestion	Biogas	[22]
7	Municipal WAS	14,000 kJ/kg TS (SE)	Anaerobic digestion	Biogas	[23]
8	Microalgae	MW power 150 W; 10 min	Fermentation	Ethanol	[24]
9	Plant biomass	MW power 560 W; 7 min	Anaerobic digestion	Biogas	[25]
10	Municipal WAS	18,000 kJ/kg TS	Anaerobic digestion	Biogas	[20]

WAS—Waste activated sludge; IE—Irradiation energy; TS—Total solids; SE—Specific energy; MW—Microwave

- Disruption of the lignin matrix to increase enzyme access to hemicellulose.
- Reduction of cellulose crystallinity.
- Increase in the surface area.
- Increase cell wall permeability in biomass.

4.3.1 Acid Pretreatment

In acid pretreatment, dried biomass is size reduced, pre-soaked in water, and immersed in an acidic solution under particular temperatures. In an acidic environment, fast hydrolysis of the hemicelluloses occurs in lignocellulosic biomass. Usually, the concentration of hydrogen ions is directly proportional to the hydrolysis rate constant. Table 4.4 lists the usage of acid in pretreating biomass. Therefore, rapid hydrolysis occurs when the pKa value of acid is more negative [26]. Widely used acids are sulphuric (H_2SO_4) and phosphoric (H_3PO_4) acids due to their low cost and their effectiveness in hydrolyzing lignocellulose. Hydrochloric (HCl) acid is more volatile and simple to regain, and hydrolysis of biomass occurs better than H_2SO_4. Similarly, nitric acid (HNO_3) increases better cellulose to sugar conversion rates.

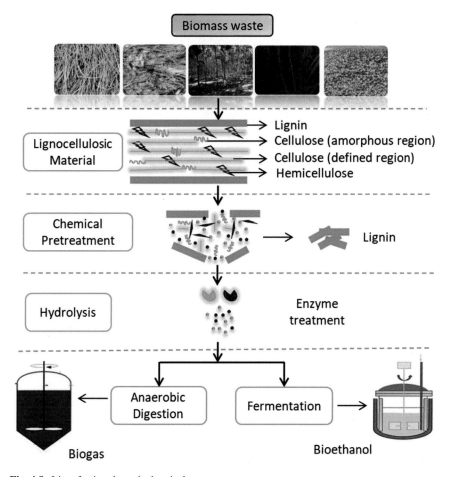

Fig. 4.8 Liquefaction through chemical process

For effective pretreatment, high temperature and pressure along with acid pretreatment are required [27]. Biomass pretreatment with dilute acid is the very effective method than the concentrated acid pretreatments for lignocelluloses, due to the low generation of degradation products.

4.3.2 Alkaline Pretreatment

Alkalis such as hydroxides of sodium, calcium, potassium, and ammonia are commonly used to pretreat biomass. Alkaline pretreatment removes tge acetyl group from hemicellulose, which in turn increases the availability of cellulosic compounds to enzyme hydrolysis. Saphonication and salvation occur, during the alkaline pretreatment [31, 32]. As a result of saponification, the inner surface area of tissue gets

Table 4.4 Usage of acid in pretreating biomass

S. No.	Biomass used	Pretreatment condition	Conversion process	Energy produced	References
1	Corn stalk	0.2% HCl	Fermentation	Biohydrogen	[28]
2	Microalgae	3% sulphuric acid (v/v)	Fermentation	Ethanol	[29]
3	Poplar leaves	4% HCl	Fermentation	Biohydrogen	[26]
4	Microalgae	1% sulphuric acid (v/v)	Fermentation	Ethanol	[27]
5	Sweet sorghum	0.5% (w/v) sulphuric acid	Fermentation	Biohydrogen	[30]
6	Oat straw	2% HCl	Fermentation	Biohydrogen	[31]
7	Grass	4% HCl	Fermentation	Biohydrogen	[32]
8	Rice hulls	0.3% (w/v) sulphuric acid	Fermentation	Ethanol	[33]
9	Microalgae	1% sulphuric acid	Fermentation	Butanol	[34]
10	Rice straw	1% sulphuric acid	Fermentation	Ethanol	[35]

increased, the extent of polymerization and crystallinity gets decreased, the structural arrangement of lignin gets distracted, and the structural linkage between lignin and polysaccharide gets disturbed. During alkaline pretreatment, ester bonds connecting cellulose and lignin are broken down, which in turn prevents the disintegration of hemicellulose; therefore, among the chemical pretreatments, this treatment is considered as the most effective method. Table 4.5 lists the usage of alkali in pretreating biomass.

The digestibility of hardwood increased from 24 to 55% and lignin got reduced to 20% during alkali pretreatment. Calcium hydroxide is one of the efficient alkali that helps in removing acetyl groups of cellulosic structures and decreases the enzyme interruption which in turn boosts the digestibility of cellulose. When straws of wheat and sorghum are pretreated in mild alkaline condition, a high rate of enzymatic saccharification occurs subsequently followed by high ethanol formation.

Among the alkali used, NaOH was found to be best for the generation of biogas from waste activated sludge. Figure 4.9 shows the effect of alkali pretreatment on sludge biomass. About 43.5% COD solubilization was achieved with 7 g/L NaOH. Excessive addition of NaOH causes a thermal effect affecting the solubilization. However, a major limitation of alkaline pretreatment was the conversion of irretrievable salt and its amalgamation inside the biomass. The management of accumulated salt has become a difficult concern in biomass pretreated by alkali.

Table 4.5 Usage of alkali in pretreating biomass

S. No.	Biomass used	Pretreatment condition	Conversion process	Energy produced	References
1	Municipal solid waste	NaOH-5 g/Kg of solid waste	Anaerobic digestion	Biogas	[31]
2	Corn stover	CaOH-1%	Fermentation	Bioethanol	[32]
3	Food waste	0.3 g NaOH/g TS	Anaerobic digestion	Methane	[36]
4	Municipal sludge	pH 11; 1 N NaOH; 24 h	Fermentation	Hydrogen	[37]
5	Sugarcane bagasse	0.04 g CaOH/g bagasse	Fermentation	Bioethanol	[38]
6	Sugarcane bagasse	2% NaOH	Fermentation	Bioethanol	[39]
7	Areca nut husk	CaOH 0.5 wt 5 of biomass	Fermentation	Bioethanol	[40]
8	Wheat straw	KOH-1%	Fermentation	Bioethanol	[41]
9	Rye straw	KOH-1%	Fermentation	Bioethanol	[42]
10	Microalgae	NaOH-4%, KOH-4%, Ca(OH)$_2$-1%	Fermentation	Bioethanol	[43]

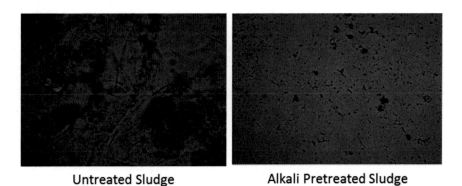

Untreated Sludge   **Alkali Pretreated Sludge**

Fig. 4.9 Effect of alkali on sludge

4.3.3 Organosolv Pretreatment

Organic solvent or organosolv pretreatment is a type of chemical pretreatment that delignifies the biomass using organic solvent or solvent mixtures with water prior to enzymatic hydrolysis. The organic solvents used for this type of pretreatment include methanol, ethanol, ethyl acetate, and aqueous acetic acid [37]. Figure 4.10 shows the process of organosolv pretreatment. This type of pretreatment involves the use of 1:1

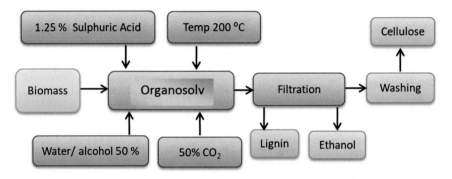

Fig. 4.10 Liquefaction through organosolv process

alcohol and carbon dioxide. This mixture solubilizes the bonds of lignin and carbohydrates and results in the formation of residual solids composed of cellulose and hemicellulose. In addition to lignin removal, this pretreatment also hydrolyzes hemicellulose. Sulphuric acid is the frequently used catalyst in organosolv pretreatment process owing to its effective reactivity and high efficiency. A higher temperature of about 200 °C and a time of 60 min can be employed for this pretreatment.

In some cases, a low temperature is enough and its selection is based on the biomass type and catalyst used. Inorganic and organic acids can be used as catalyst for this process. The solvent mixtures used for this type of pretreatment can hamper the subsequent fermentation process. So before fermentation, the solvent must be removed through filtration and washing to avoid inhibition. The advantages of organosolv pretreatment were as follows: the enzyme cost can be minimized through delignification, and as a result of this, the amenability of cellulosic structure to hydrolysis process is enhanced. Organosolv pretreatment can also minimize the adsorption of enzymes on the surface of lignin by removing it.

4.4   Oxidative Delignification

Lignocellulosic biomass can be pretreated with oxidizing agents such as hydrogen peroxide and ozone for the removal of lignin from biomass. The aromatic ring structures of lignin are destroyed by highly reactive oxidizing agents. As a result, the polymeric lignin will be transformed into organic acids such as carboxylic acids. The formed acids are inhibitory to the fermentation process. Therefore, it has to be removed or neutralized. An oxidizing agent that can be normally used is hydrogen peroxide (H_2O_2). A 50% liquefaction of lignin and hemicellulose can be obtained by pretreating the biomass with hydrogen peroxide at 27 °C [44]. As a result, the outcome of the hydrolysis process can be increased up to 95%. The major disadvantage of oxidative delignification is it destroys the hemicellulose fraction of the lignocellulose complex.

4.5 Ozone Pretreatment

Pretreatment of sludge through ozone was one of the potent technologies with the highest disintegration capability. Ozone pretreatment is a type of oxidative chemical pretreatment for disintegrating organic biomass. The experimental setup of ozone pretreatment was illustrated in Fig. 4.11. It disrupts the lignin of lignocellulosic biomass by breaking and destroying the aromatic rings arrangement, whereas hemicellulose and cellulose are barely degraded.

Ozone penetrates the cell wall of biomass, damages the cell membrane, and releases the intracellular components into the aqueous medium. This pretreatment can effectively break the structure of various lignocellulosic biomass such as wheat and cotton residues, sugarcane bagasse, groundnut, cotton straw, and sawdust [45]. Table 4.6 lists the usage of ozone in pretreating biomass. There are three potential pathways by which the ozone attacks the sludge biomass. At first, ozone deagglomerates the sludge biomass by solubilizing the extracellular polymeric substances and sludge flocs. Secondly, the ozone penetrates the cell wall of sludge biomass and releases the cell contents into the aqueous phase. Finally, ozone mineralizes the released intracellular components.

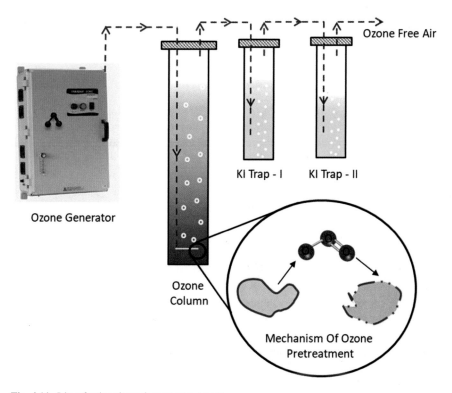

Fig. 4.11 Liquefaction through ozone treatment

Table 4.6 Usage of ozone in pretreating biomass

S. No.	Biomass used	Pretreatment condition	Conversion process	Energy produced	References
1	Canned maize sludge	Contact time 30 min	Anaerobic digestion	Biogas	[47]
2	Municipal WAS	0.15 g O_3/g TS	Anaerobic digestion	Methane	[48]
3	Plant biomass	Contact time 30–90 min	Fermentation	Ethanol	[49]
4	Municipal WAS	0.05–0.5 gO_3/gTS	Anaerobic digestion	Biogas	[50]
5	Municipal solid waste	0.16 gO_3/gTS	Anaerobic digestion	Methane	[51]
6	Sugarcane bagasse	Contact time 60 min	Fermentation	Ethanol	[52]
7	Dairy WAS	0.0011 mgO_3/mg SS	Anaerobic digestion	Biogas	[53]
8	Sewage sludge	0.012 gO_3/g TS	Anaerobic digestion	Biogas	[54]
9	Municipal WAS	0.036 g O_3/g TS	Anaerobic digestion	Biogas	[55]
10	Microalgae	382 mg O^3 g^{-1} VS	Anaerobic digestion	Methane	[56]

WAS—Waste activated sludge; SS—Suspended solids; TS—Total solids; O_3—Ozone

The application of microbubble ozonation accelerates the formation of hydroxyl radicals and speeds up sludge solubilization, thus reducing the impact of high capital requirements. About 0.05–0.5 g O_3/g TS enhances sludge solubilization. Ozonation consumes higher energy of 12.5 kWh/kg O_3. To reduce this energy loss, microbe ozonation was developed. Using this method, solubilization increases from 15 to 30% at the dose of 0.06–0.16 g O_3/g TSS [46]. The microbubbles generated involved in the generation of hydroxyl radical stimulating better solubilization.

4.6 Biological Pretreatment

In biological pretreatment, metabolic activity (capable of producing enzymes) of microorganisms (fungi and bacteria) degrades the cell wall of biomass and enhances biofuel production. Biological pretreatment has numerous advantages like eco-friendly, no fermentation inhibitors, and enhanced enzyme activity which led to economically viable approach. In addition, the process does not demand the usage of chemicals and it does not discharge toxic substances into the environment. Biological pretreatment involving the usage of white rot fungi can hydrolyze lignin and appears

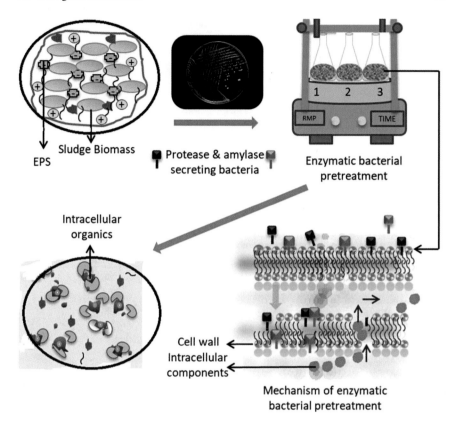

Fig. 4.12 Liquefaction through biological process

to be a favourable method due to its less energy consumption and minimal harm to the environment [57].

During biological pretreatment, microbes produce many types of extracellular hydrolytic and oxidative enzymes. Several white rot fungi such as Phanerochaetechrysosporium, Ceriporia lacerate, Cyathusstercolerus, Ceriporiopsissubvermispora, Pycnoporuscinnarrbarinus, and Pleurotusostreaus have exhibited great delignification effectiveness on various lignocellulosic biomass. Figure 4.12 shows the schematic representation of cell wall cleavage induced by biological pretreatment.

The lytic enzyme plays a key role in overcoming the limitation of the hydrolysis step in the biodegradation process. Enzymatic pretreatment of activated sludge biomass for bioenergy generation demands more cost. Therefore, cultivating microorganism secreting hydrolytic enzyme became an efficient approach. The hydrolytic enzyme, lysozyme has the capability of degrading bacterial cell wall, which consists of polysaccharides. Lysozyme secreting bacteria such as Bacillus jerish 03 and jerish 04 can disintegrate sludge biomass and enhance biofuel production. Protease has the potential of hydrolyzing the peptide bonds of proteins into peptides and amino

Table 4.7 Usage of biological disintegration in pretreating biomass

S. No.	Biomass used	Pretreatment conditions	Conversion process	Energy produced	References
1	Microalgae	0.2% glucoamylase	Fermentation	Bioethanol	[45]
2	Microalgae	0.4 mL g^{-1} of amyloglucosidase	Fermentation	Bioethanol	[46]
3	Municipal WAS	77 × 106 bacterial cells mg^{-1} of SS	Anaerobic digestion	Methane	[59]
4	Municipal WAS	2 g dry cell weight/L	Anaerobic digestion	Biogas	[60]
5	Municipal WAS	69 × 10^7 CFU/mL	Anaerobic digestion	Biogas	[60]
6	Municipal WAS	2.5 g dry cell weight/L	Anaerobic digestion	Methane	[25]
7	Dairy WAS	2 g dry cell weight/L	Anaerobic digestion	Methane	[32]
8	Municipal WAS	0.66 g dry cell weight/L	Anaerobic digestion	Methane	[61]
9	Microalgae	0.7 g dry cell weight/L	Anaerobic digestion	Methane	[62]
10	Microalgae	0.52 g dry cell weight/L	Anaerobic digestion	Methane	[35]

WAS—Waste activated sludge; SS—Suspended solids

acids, which significantly increases the effectiveness of biomass hydrolysis. Protease enzyme has the ability to procatalyze peptide bond and enhances sludge degradation. The exoenzyme secreting bacteria could induce 20–23% of COD solubilization when it was mediated by the chemicals. The biosurfactant secreting bacteria achieved 20–22% of COD solubilization, and protease secreting bacteria undergo COD solubilization greater than 20% [58]. Hydrolytic enzymes such as protease, amylase and cellulase have the capacity to improve the effectiveness of sludge disintegration. Cellulolytic, proteolytic, and amylolytic bacteria have been used for disintegrating microalgal biomass and achieved a higher energy ratio during biomethanation of microalgae (Chlorella vulgaris).

Biological pretreatment has few set back as the added microorganisms are not always selective. For example, lignolytic microorganisms in addition to lignin removal also hydrolyze cellulosic structures and this makes it commercially less attractive. Table 4.7 lists the usage of biological disintegration in pretreating biomass.

4.7 Steam Explosion

Steam explosion is the most commonly employed pretreatment method for ligno-cellulosic biomass. It is highly popular among other pretreatments due to its less chemical usage and energy requirement. In this type of pretreatment, both physical and chemical methods (hydrothermal pretreatment) are employed to disintegrate lignocellulosic biomass. The biomass to be pretreated is subjected to high-pressure saturated steam. As a result of this, the temperature of the medium rises to 160–260 °C. Subsequent to this, the pressure of the medium is minimized and biomass undergoes explosive decompression [63–65]. As a result of this decompression, the fibril structure of biomass will get destroyed (hemicellulose degradation and lignin matrix destruction).

The defibration considerably increases the substrate degradability, bioconversion, and bioenergy generation. Chemicals such as acids and alkali can be used in combination with the steam explosion. During the steam explosion, some inhibitory compounds are produced which may hamper the downstream processes. These products must be separated before the fermentation process. The factors that influence the pretreatment are treatment time, temperature, size of biomass, and moisture. Figure 4.13 shows the steam explosion unit.

4.8 Wet Oxidation

Wet oxidation is a type of physiochemical pretreatment process in which oxygen combines with water at high temperature and pressure to disintegrate the biomass. It can be considered as an alternative to steam explosion pretreatment. It is employed to disintegrate biomass having high organic content. During wet oxidation, the biomass is liquefied in oxygenic conditions at elevated temperatures (150–350 °C) and pressure (5–20 MPa). Wet oxidation is mainly employed for pretreating biomass such as wheat straw and hardwood. During wet oxidation, wheat straw was subjected to an alkaline environment. The main products formed were carboxylic acids, carbon dioxide, and water and which come from the degradation of cellulose and hemicellulose. In comparison with other oxidative pretreatments, wet oxidation is highly effective in pretreating lignocellulosic biomass since it breaks the crystalline structure of cellulose. Through wet oxidation of wheat straw, approximately 65% degree of delignification can be achieved. In the case of woody biomass, wet oxidation pretreatment has been proven to liquefy largely the hemicellulose portion [66]. The main benefit of this pretreatment is it produces a minimal quantity of inhibitor compounds such as furfural and 5-hydroxymethylfurfural compounds. These compounds are toxic to the fermentation process.

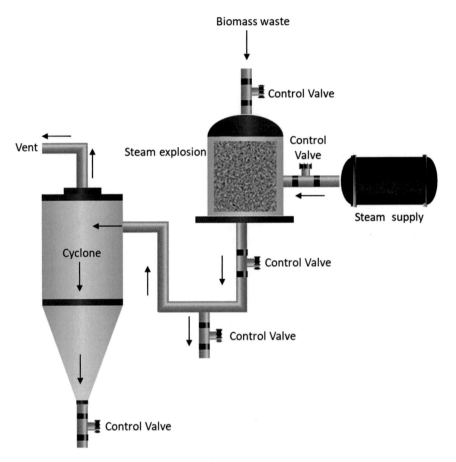

Fig. 4.13 Liquefaction through steam explosion

4.9 Advanced Oxidation Pretreatment

TiO_2, ZnO, CuO, SnO_2, and other photocatalysts are used in photocatalytic pretreatment. This photocatalyst becomes active when exposed to UV and solar radiation which results in hydroxyl radicals generation. Photocatalysts may be recovered and reused, which is a major advantage of photocatalytic pretreatment. Solubilization is increased by more than 20% when advanced oxidation is used as a pretreatment. The pretreatment effectiveness of TiO_2 achieved a SCOD/TCOD ratio of 0.92 at a TiO_2 dose of 5 mg/L when exposed to UV light for 12 h. The challenge of retrieving TiO_2 can be minimized by immobilizing the TiO_2 particle on a substrate. Nearly, 18% COD solubilization is obtained by immobilizing TiO_2 on a glass platter [67]. Also, 60 g H_2O_2/kg TS and 0.07 g Fe^{2+}/g H_2O_2, respectively, can be used to achieve better solubilization in sludge biomass. This photo-Fenton under solar light achieves higher solubilization with the organic release of 4.5 times greater than the initial COD and 40% of sludge gets reduced.

4.10 Combinative Pretreatment

Each pretreatment has its own unique disintegration mechanisms to liquefy biomass. Therefore, pretreatment with different disintegration mechanisms can be combined to achieve an increase in disintegration efficiency of biomass in cost-effective manner. Combinative pretreatment is economically more beneficial than individual pretreatment in disintegrating biomass. There are various combinative pretreatment methods such as physiochemical (steam explosion, wet oxidation), chemo-mechanical, and biomechanical pretreatments. Combined chemo-biological pretreatment is an efficient combinative pretreatment for better lignin removal and enhanced bioethanol generation in plant biomass, water hyacinth. Pretreatment with diluted acid at moderate temperatures breaks the lignin which in turn liberates hemicellulose and cellulose and makes it available for enzymatic action. Hydrothermal pretreatment in combination with alkali enhances the breakdown of lignin by means of ester bond cleavage. Enhancement of enzyme hydrolysis occurs in oil palm mesocarp fibre by combined hydrothermal, alkaline, and mechanical pretreatment. Through combined thermochemo ozone pretreatment, Kannah et al. [68] have achieved better results than the individual pretreatment method. From Fig. 4.14, it was evident that a combination of pretreatment techniques yielded a greater energy ratio (0.61) than individual pretreatment (0.11). Figure 4.14 shows the effect of thermal and thermochemo ozone on energy ratio during the pretreatment of waste activated sludge biomass. Combinative process not only increases energy ratio but also decreases the cost required to pretreat the biomass. Table 4.8 lists the usage of combinative disintegration in pretreating biomass. During high-temperature thermochemical pretreatment, the sludge solubilization increased from 28 to 87%. It reduces chemical consumption 4–6 times less than thermochemical pretreatment. Adding NaOH at 121 °C sludge promotes 87% COD solubilization.

Nearly 50–70% COD solubilization was achieved during sono-chemo treatment. Mechano-chemo treatment induces 21% of COD solubilization. Biogas output was enhanced four-fold times which is nearly from 0.31 to 1.14 m^3 biogas/kgVS during pasteurization of animals by-products (ABP).

4.11 Evaluation of Pretreatment

Evaluation of pretreatment is essential; otherwise, it will increase the input energy and the cost of the process. In some cases, it leads to the accumulation of inhibitors or toxic components in the medium. This in turn reduces the efficiency of biofuel production. Hence, it is important to evaluate the extent of pretreatment for cost-effective biofuel production. The methods available to evaluate the efficiency of pretreatment include the following:

- Hydrolysis,
- Biochemical methane potential, and
- Energy balance.

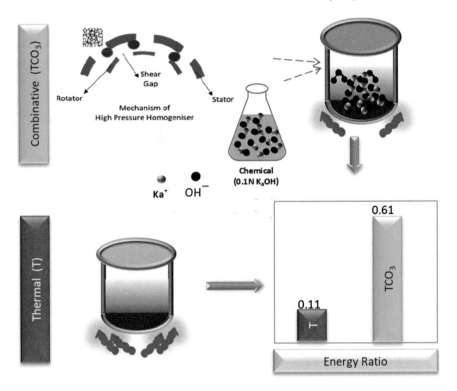

Fig. 4.14 Single and combinative liquefaction process

Biochemical methane potential (BMP)

BMP assay is the test performed to evaluate the efficiency of pretreatment on biomethane production. This assay is used to estimate the amount of organic compound (usually estimated as chemical oxygen demand) of a given substrate released during pretreatment that can be anaerobically converted to methane. In BMP assay, the solubilization potential of a biomass is considered to be the main factor that influences its biomethane production. The hydrolysis rate of complex particulate organics is lower when compared to simple dissolvable organics since the smaller the particles, the larger the available surface area for the action of methanogens. This in turn results in better methane generation. During pretreatment, water-insoluble complex organic matter gets converted into soluble simple monomers. The more the availability of soluble monomers (simple carbohydrates, proteins, and simple lipids), the faster the fermentation. This in turn enhances the biomethane production. The experimental setup of the BMP assay is illustrated in Fig. 4.15.

Modelling is an essential tool for assessing the bioprocess and comparing the methane production potential of different pretreated biomass with that of untreated biomass. Anaerobic biodegradation is a complex bioprocess. Various models have been reported in many literature for parameter identification and validation of a

Table 4.8 Usage of combinative disintegration in pretreating biomass

S. No.	Biomass used	Pretreatment conditions	Conversion process	Energy produced	References
1	Microalgae	NaOH-0.75%; temperature-120 °C for 30 min (thermochemical)	Fermentation	Bioethanol	[32]
2	Dairy WAS	NaOH-pH 12; Temp 60 °C	Anaerobic digestion	Biogas	[60]
3	Dairy WAS	4172 kJ/kg TS (SE); NaOH-pH 10 (Sonic + alkali)	Anaerobic digestion	Methane	[16]
4	Municipal WAS	3360.94 kJ/kg TS; NaOH-pH 10; 80 °C (thermochemo disperser)	Anaerobic digestion	Methane	[21]
5	Municipal WAS	5290.5 kJ/kg TS; NaOH, KOH and Ca(OH)$_2$ pH 11; 80 °C (thermochemo sonic)	Anaerobic digestion	Methane	[61]
6	Municipal WAS	50 mM citric acid; 2 g dry cell weight (g/L) (Bacteria + cation binding agent)	Anaerobic digestion	Methane	[69]
7	Dairy WAS	18,600 kJ/kg TS; 0.3 mg/g SS of H$_2$O$_2$; acidic pH 5 (Microwave + H$_2$O$_2$ + acid)	Anaerobic digestion	Biogas	[70]
8	Municipal WAS	5500 kJ/ kg TS; NaOH-pH 11; 80 °C (thermochemo sonic)	Anaerobic digestion	Methane	[71]
9	Municipal WAS	NaOH-pH 11; 80 °C; ozone-0.0012 mg O$_3$/mg SS; SE-141.02 kJ/kg TS; (thermochemo ozone)	Anaerobic digestion	Methane	[68]
10	Macroalgae	3312.6 kJ/kg TS; Surfactant STPP-0.04 g/g COD; (disperser + surfactant)	Anaerobic digestion	Methane	[29]

WAS—Waste activated sludge; SE—Specific energy; TS—Total solids; SE—Specific energy; SS—Suspended solids

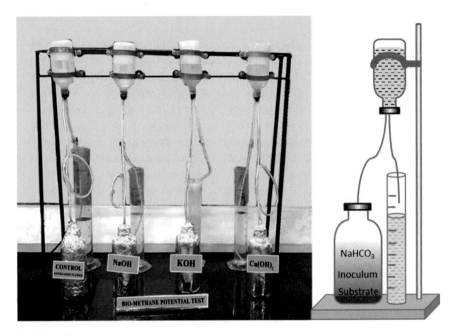

Fig. 4.15 BMP experimental setup

Table 4.9 Biomethane production potential of different pretreatments

S. No.	Substrate	Pretreatment	Biomethane production potential	References
1	Municipal biosolids	Bacterial	0.34 (L/g VS)	[72]
2	Municipal biosolids	Microwave	0.38(L/g VS)	[17]
3	Dairy biosolids	Disperser	1.1 (L/g VS)	[69]
4	Algal biomass	Ultrasound	0.28(L/g VS)	[73]
5	Plant biomass	Microwave	0.078 (L/g TS)	[74]

pretreatment process on biomethane production. Among the models, the Gompertz model is widely used due to its simplicity and accuracy. It was used to arrive at anaerobic kinetic parameters of substrates. Table 4.9 shows the kinetic parameters of the biomethane production potential of different pretreatments. Biomethane production potential was evaluated through the Gompertz equation as follows:

$$(Bio)T = Bio * \exp\left[-\exp\left[RBio / Bio * \exp(\lambda - T) + 1\right]\right] \qquad (4.3)$$

where

  $(Bio)T$ = Total amount of biogas production at time (T),
  Bio = Biomethane generation potential (L/(g VS)),
  $RBio$ = Maximal biomethane generation rate (L/(g VS d)), and

λ = Lag phase (days).

4.12   Energy Balance Assessment

The bioenergy generation potential of biomass pretreatment can be evaluated through an anaerobic biodegradability assay in the laboratory. However, energy balance is essential for assessing the economic feasibility of implementing a pretreatment process at pilot-scale level.

The energy spent must be balanced by the energy generated in order to make a pretreatment process feasible. Energy analysis was carried out by mass balancing total energy input and output. The outcome of energy balance analysis can be articulated in two modes: net energy production and energy ratio.

Net energy production: The energy balance between the output (OE) and the input energy (IE), and it can be expressed as follows:

$$\bullet \quad \text{Net energy production} = \text{OE} - \text{IE} \qquad (4.4)$$

Energy ratio: The energy ratio between the OE and IE, and it can be expressed as follows:

- Energy ratio = OE/IE.

Table 4.10 summarizes the energy ratio of different pretreatments. The field applicability of a pretreatment process is based on its energy ratio. Pretreatment processes resulting in an energy ratio greater than one are scalable.

Advantages of pretreatment

The following advantages are associated with the pretreatment of biomass:

- Increase the availability of surface area.
- Lessen the crystallization of cellulose.
- Hydrolysis of hemicelluloses and lignin.
- Breakdown of the lignin structure.
- Improve the enzymatic accessibility to the maximum.
- Loss of sugars can be minimized.

Table 4.10 Energy ratio of different pretreatments

S. No.	Pretreatment	Energy ratio	References
1	Thermal	0.59	[75]
2	Hydrothermal	7.78	[17]
3	Microwave	0.05	[76]
4	Ultrasound	1.00	[77]
5	Biological	1.04	[78]

- High disintegration efficiency leads to high biogas yield.
- Make the anaerobic process relatively simple.
- Increase anaerobic process stability.
- Degradation of hard degradable substances.
- Decrease of capital and operating costs.
- A well-organized pretreatment should preserve the fractions of hemicellulose and lessen the formation of toxic substances, which in turn prevents the growth of fermentative microorganisms.

Disadvantages of pretreatment

The following disadvantages are associated with the pretreatment of biomass:

- Formation of recalcitrant substances during the pretreatment of biomass.
- Increase the cost of biofuel production.

References

1. Ma J, Frear C, Wang Z-W, Yu L, Zhao Q, Li X, Chen S (2013) A simple methodology for rate-limiting step determination for anaerobic digestion of complex substrates and effect of microbial community ratio. Bioresour Technol 134:391–395
2. Mairet F, Bernard O, Masci P, Lacour T, Sciandra A (2011) Modelling neutral lipid production by the microalga Isochrysis aff. galbana under nitrogen limitation. Bioresour Technol 102:142–149
3. Kythreotou N, Florides G, Tassou SA (2014) A review of simple to scientific models for anaerobic digestion. Renew Energy 71:701–714
4. Kafle GK, Kim SH, Sung KI (2013) Ensiling of fish industry waste for biogas production: a lab scale evaluation of biochemical methane potential (BMP) and kinetics. Bioresour Technol 127:326–336
5. Mejdoub H, Ksibi H (2015) Regulation of biogas production through waste water anaerobic digestion process: modeling and parameters optimization. Waste Biomass Valorizat 6:29–35
6. Liang Y, Lu Y, Li Q (2016) Comparative study on the performances and bacterial diversity from anaerobic digestion and aerobic composting in treating solid organic wastes
7. Patil JH, Raj MA, Muralidhara PL, Desai SM, Raju GKM (2012) Kinetics of anaerobic digestion of water hyacinth using poultry litter as inoculum. Int J Environ Sci Dev 94–98
8. Ganidi N, Tyrrel S, Cartmell E (2009) Anaerobic digestion foaming causes–a review. Bioresour Technol 100:5546–5554
9. Chen H, Yan S-H, Ye Z-L, Meng H-J, Zhu Y-G (2012) Utilization of urban sewage sludge: Chinese perspectives. Environ Sci Pollut Res 19:1454–1463
10. Perendeci A, Arslan S, Celebi S, Tanyolac A (2008) Prediction of effluent quality of an anaerobic treatment plant under unsteady state through ANFIS modeling with on-line input variables. Chem Eng J 145:78–85
11. Qdais HA, Hani KB, Shatnawi N (2010) Modeling and optimization of biogas production from a waste digester using artificial neural network and genetic algorithm. Resour Conserv Recycl 54:359–363
12. Shen S, Premier GC, Guwy A, Dinsdale R (2007) Bifurcation and stability analysis of an anaerobic digestion model. Nonlinear Dyn 48:391–408
13. Siegrist H, Vogt D, Garcia-Heras JL, Gujer W (2002) Mathematical model for meso- and thermophilic anaerobic sewage sludge digestion. Environ Sci Technol 36:1113–1123

14. Subha C, Kavitha S, Abisheka S, Tamilarasan K, Arulazhagan P, Banu JR (2019) Bioelectricity generation and effect studies from organic rich chocolaterie wastewater using continuous upflow anaerobic microbial fuel cell. Fuel 251:224–232

15. Blumensaat F, Keller J (2005) Modelling of two-stage anaerobic digestion using the IWA anaerobic digestion model no. 1 (ADM1). Water Res 39:171–183

16. Wang P, Yu Z, Zhao J, Zhang H (2018) Do microbial communities in an anaerobic bioreactor change with continuous feeding sludge into a full-scale anaerobic digestion system? Bioresour Technol 249:89–98

17. Preethi BJR, Sharmila VG, Kavitha S, Varjani S, Kumar G, Gunasekaran M (2021) Alkali activated persulfate mediated extracellular organic release on enzyme secreting bacterial pretreatment for efficient hydrogen production. Bioresour Technol 341:125810

18. Sun C, Cao W, Liu R (2015) Kinetics of methane production from swine manure and buffalo manure. Appl Biochem Biotechnol 177:985–995

19. Taricska JR, Long DA, Chen JP, Hung Y-T, Zou S-W (2007) Anaerobic digestion. In: Handbook of environmental engineering. Humana Press, Totowa, NJ, pp 135–176

20. Bhatia SK, Otari SV, Jeon J-M, Gurav R, Choi Y-K, Bhatia RK, Pugazhendhi A, Kumar V, Banu JR, Yoon J-J, Choi K-Y, Yang Y-H (2021) Biowaste-to-bioplastic (polyhydroxyalkanoates): conversion technologies, strategies, challenges, and perspective. Bioresour Technol 326:124733

21. Puyuelo B, Gea T, Sánchez A (2010) A new control strategy for the composting process based on the oxygen uptake rate. Chem Eng J 165:161–169

22. Behera SK, Meher SK, Park H-S (2015) Artificial neural network model for predicting methane percentage in biogas recovered from a landfill upon injection of liquid organic waste. Clean Technol Environ Policy 17:443–453

23. Banu JR, Sharmila VG, Ushani U, Amudha V, Kumar G (2020) Impervious and influence in the liquid fuel production from municipal plastic waste through thermo-chemical biomass conversion technologies—a review. Sci Total Environ 718:137287

24. Banu JR, Preethi KS, Tyagi VK, Gunasekaran M, Karthikeyan OP, Kumar G (2021) Lignocellulosic biomass based biorefinery: a successful platform towards circular bioeconomy. Fuel 302:121086

25. Ginni K, Kannah RY, Bhatia SK, Kumar A, Rajkumar KG, Pugazhendhi A, Chi NTL, Banu R (2021) Valorization of agricultural residues: Different biorefinery routes. J Environ Chem Eng 9:105435

26. Terashima M, Goel R, Komatsu K, Yasui H, Takahashi H, Li YY, Noike T (2009) CFD simulation of mixing in anaerobic digesters. Bioresour Technol 100:2228–2233

27. Vavilin VA, Lokshina LY, Flotats X, Angelidaki I (2007) Anaerobic digestion of solid material: multidimensional modeling of continuous-flow reactor with non-uniform influent concentration distributions. Biotechnol Bioeng 97:354–366

28. Sharmila VG, Kumar SA, Banu JR, Yeom IT, Saratale GD (2019) Feasibility analysis of homogenizer coupled solar photo Fenton process for waste activated sludge reduction. J Environ Manage 238:251–256

29. Di Maria F, Barratta M, Bianconi F, Placidi P, Passeri D (2017) Solid anaerobic digestion batch with liquid digestate recirculation and wet anaerobic digestion of organic waste: Comparison of system performances and identification of microbial guilds. Waste Manag 59:172–180

30. Madigan MT, Martinko JM (2006) Brock biology of microorganisms, 11th edn. Prentice Hall, Pearson

31. Vavilin VA, Fernandez B, Palatsi J, Flotats X (2008) Hydrolysis kinetics in anaerobic degradation of particulate organic material: an overview. Waste Manag 28:939–951

32. Vlyssides A, Barampouti EM, Mai S (2007) An alternative approach of UASB dynamic modeling. AIChE J 53:3269–3276

33. Raj T, Chandrasekhar K, Banu R, Yoon J-J, Kumar G, Kim S-H (2021) Synthesis of γ-valerolactone (GVL) and their applications for lignocellulosic deconstruction for sustainable green biorefineries. Fuel 303:121333

34. Banu JR, Tamilarasan K, Chang SW, Nguyen DD, Ponnusamy VK, Kumar G (2020) Surfactant assisted microwave disintegration of green marine macroalgae for enhanced anaerobic biodegradability and biomethane recovery. Fuel (Lond) 281:118802

35. Ahn HK, Richard TL, Choi HL (2007) Mass and thermal balance during composting of a poultry manure Wood shavings mixture at different aeration rates. Process Biochem 42:215–223

36. Chang JI, Chen YJ (2010) Effects of bulking agents on food waste composting. Bioresour Technol 101:5917–5924

37. Wu B (2012) Integration of mixing, heat transfer, and biochemical reaction kinetics in anaerobic methane fermentation. Biotechnol Bioeng 109:2864–2874

38. Wei L, Shutao W, Jin Z, Tong X (2014) Biochar influences the microbial community structure during tomato stalk composting with chicken manure. Bioresour Technol 154:148–154

39. Chynoweth DP, Turick CE, Owens JM, Jerger DE, Peck MW (1993) Biochemical methane potential of biomass and waste feedstocks. Biomass Bioenerg 5:95–111

40. Pugazhendi A, Alreeshi GG, Jamal MT, Karuppiah T, Jeyakumar RB (2021) Bioenergy production and treatment of aquaculture wastewater using saline anode microbial fuel cell under saline condition. Environ Technol Innov 21:101331

41. Wu B (2012) CFD simulation of mixing for high-solids anaerobic digestion. Biotechnol Bioeng 109:2116–2126

42. Banu JR, Kumar MD, Gunasekaran M, Kumar G (2019) Biopolymer production in bio electrochemical system: literature survey. Bioresour Technol Rep 7:100283

43. Gopikumar S, Banu JR, Robinson YH, Shanmuganathan V, Kadry S, Rho S (2021) Novel framework of GIS based automated monitoring process on environmental biodegradability and risk analysis using Internet of Things. Environ Res 194:110621

44. Yu L, Ma J, Chen S (2011) Numerical simulation of mechanical mixing in high solid anaerobic digester. Bioresour Technol 102:1012–1018

45. Yu L, Wensel PC (2013) Mathematical modeling in anaerobic digestion (AD). J Bioremediat Biodegrad 4:1–12

46. Neto JM, Dos Reis GD, Rueda SMG, da Costa AC (2013) Study of kinetic parameters in a mechanistic model for enzymatic hydrolysis of sugarcane bagasse subjected to different pretreatments. Bioprocess Biosyst Eng 36:1579–1590

47. Beszedes S, Kertesz SZ, Laszlo Z, Szabo G (2007) Biogas production of ozone and/or microwave-pretreated canned maize production sludge. Ozone Sci Eng 2:1–3

48. Bougrier C, Battimelli A, Delgenes JP, Carrere H (2007) Combined ozone pretreatment and anaerobic digestion for the reduction of biological sludge production in wastewater treatment. Ozone Sci Eng 29:201–206

49. Silverstein RA, Chen Y, Sharma-Shivappa RR, Boyette MD, Osborne J (2007) A comparison of chemical pretreatment methods for improving saccharification of cotton stalks. Bioresour Technol 98:3000–3011

50. Carrere H, Dumas C, Battimelli A, Batstone DJ, Delgenes JP, Steyer JP, Ferrer I (2010) Pretreatment methods to improve sludge anaerobic degradability: a review. J Hazard Mater 183:1–15

51. Cesaro A, Belgiorno V (2013) Sonolysis and ozonation as pretreatment for anaerobic digestion of solid organic waste. Ultrason Sonochem 2013(20):931–936

52. Barros Rda R, Paredes Rde S, Endo T, Bon EP, Lee SH (2013) Association of wet disk milling and ozonolysis as pretreatment for enzymatic saccharification of sugarcane bagasse and straw. Bioresour Technol 136:288–294

53. Sowmya Packyam G, Kavitha S, Kumar A, Kaliappan S, Yeom IT, Rajesh Banu J (2015) Effect of sonically induced deflocculation on the efficiency of ozone mediated partial sludge disintegration for improved production of biogas. Ultrason Sonochem 26:241–248

54. Tian X, Trzcinski AP, Lin LL, Ng WJ (2015) Impact of ozone assisted ultrasonication pretreatment on anaerobic digestibility of sewage sludge. J Environ Sci 33:29–38

55. Tian X, Wang C, Trzcinski AP, Lin LL, Ng WJ (2015) Interpreting the synergistic effect in combined ultrasonication–ozonation sewage sludge pre-treatment. Chemosphere 140:63–71

56. Cardena R, Moreno G, Bakonyi P, Buitron G (2017) Enhancement of methane production from various microalgae cultures via novel ozonation pretreatment. Chem Eng J 307:948–954
57. Saha BC, Iten LB, Cotta MA, Wu YV (2005) Dilute acid pretreatment, enzymatic saccharification and fermentation of wheat straw to ethanol. Process Biochem 40:3693–3700
58. Saeman JF (1945) Kinetics of wood saccharification—hydrolysis of cellulose and decomposition of sugars in dilute acid at high temperature. Ind Eng Chem 37:43–52
59. Rani G, Nabi Z, Banu JR, Yogalakshmi KN (2020) Batch fed single chambered microbial electrolysis cell for the treatment of landfill leachate. Renew Energy 153:168–174
60. Uthirakrishnan U, Sharmila VG, Merrylin J, Kumar SA, Dharmadhas JS, Varjani S, Banu JR (2022) Current advances and future outlook on pretreatment techniques to enhance biosolids disintegration and anaerobic digestion: a critical review. Chemosphere 288:132553
61. Sharmila VG, Gunasekaran M, Angappane S, Zhen G, Tae Yeom I, Banu JR (2019) Evaluation of photocatalytic thin film pretreatment on anaerobic degradability of exopolymer extracted biosolids for biofuel generation. Bioresour Technol 279:132–139
62. Demirel B, Scherer P (2008) The roles of acetotrophic and hydrogenotrophic methanogens during anaerobic conversion of biomass to methane: a review. Rev Environ Sci Biotechnol 7:173–190
63. Scott F, Li M, Williams DL, Conejeros R, Hodge DB, Aroca G (2015) Corn stover semi-mechanistic enzymatic hydrolysis model with tight parameter confidence intervals for model-based process design and optimization. Bioresour Technol 177:255–265
64. Tao G, Lestander TA, Geladi P, Xiong S (2012) Biomass properties in association with plant species and assortments I: a synthesis based on literature data of energy properties. Renew Sustain Energy Rev 16:3481–3506
65. Tao ZH, Wang SX, Ji LX, Zheng L, He W (2013) Electrochemical investigation of the adsorption behaviour of guanine on copper in acid medium. Adv Mat Res 787:30–34
66. Ulas S, Diwekar UM (2004) Thermodynamic uncertainties in batch processing and optimal control. Comput Chem Eng 28:2245–2258
67. Wang J, Ye J, Yin H, Feng E, Wang L (2012) Sensitivity analysis and identification of kinetic parameters in batch fermentation of glycerol. J Comput Appl Math 236:2268–2276
68. Zhang Y-HP, Lynd LR (2004) Toward an aggregated understanding of enzymatic hydrolysis of cellulose: noncomplexed cellulase systems. Biotechnol Bioeng 88:797–824
69. Liang Y, Lu Y, Li Q (2016) Comparative study on the performances and bacterial diversity from anaerobic digestion and aerobic composting in treating solid organic wastes. Waste Biomass Valor 8:425–432
70. Cirne DG, Björnsson L, Alves M, Mattiasson B (2006) Effects of bioaugmentation by an anaerobic lipolytic bacterium on anaerobic digestion of lipid-rich waste. J Chem Technol Biotechnol 81:1745–1752
71. Sharmila VG, Kavitha S, Obulisamy PK, Banu JR (2020) Production of fine chemicals from food wastes. In: Food waste to valuable resources. Elsevier, pp 163–188
72. Fu B, Jiang Q, Liu H-B, Liu H (2015) Quantification of viable but nonculturable Salmonella spp. and Shigella spp. during sludge anaerobic digestion and their reactivation during cake storage. J Appl Microbiol 119:1138–1147
73. Gill AO, Holley RA (2003) Interactive inhibition of meat spoilage and pathogenic bacteria by lysozyme, nisin and EDTA in the presence of nitrite and sodium chloride at 24 °C. Int J Food Microbiol 80:251–259
74. Roman HJ, Burgess JE, Pletschke BI (2006) Enzyme treatment to decrease solids and improve digestion of primary sewage sludge. Afr J Biotechnol 5:963–967
75. Das KC (2008) Co-composting of alkaline tissue digester effluent with yard trimmings. Waste Manag 28:1785–1790
76. Pugazhendi A, Al-Mutairi AE, Jamal MT, Banu JR, Palanisamy K (2020) Treatment of seafood industrial wastewater coupled with electricity production using air cathode microbial fuel cell under saline condition. Int J Energy Res 44:12535–12545
77. Sánchez-García M, Alburquerque JA, Sánchez-Monedero MA, Roig A, Cayuela ML (2015) Biochar accelerates organic matter degradation and enhances N mineralisation during

composting of poultry manure without a relevant impact on gas emissions. Bioresour Technol 192:272–279

78. Bhatia SK, Jagtap SS, Bedekar AA, Bhatia RK, Patel AK, Pant D, Banu JR, Rao CV, Kim Y-G, Yang Y-H (2020) Recent developments in pretreatment technologies on lignocellulosic biomass: effect of key parameters, technological improvements, and challenges. Bioresour Technol 300:122724

Chapter 5
Kinetics and Modelling of Hydrolysis

Introduction

Understanding the process of hydrolysis is very vital as it is a precursor to the anaerobic digestion process wherein the generation of biofuel occurs. Ineffective hydrolysis process leads to a reduction in biofuel generation and makes the whole process of digestion futile and uneconomic. To assess the process, it is necessary that the mechanism of hydrolysis and all parameters involved in the process such as kinetic is need to be understood and determined. This chapter discusses exclusively the kinetics and modelling part, so that it will help in the design of reactors and model the generation of biofuels.

Kinetics

Hydrolysis rates are generally first-order or *pseudo* first-order under most environmental conditions with an overall observed hydrolysis rate constant kh. The half-life can therefore be expressed as

$$t_{1/2} = \frac{\ln 2}{kh} \tag{5.1}$$

Hydrolysis reactions are generally enhanced by both acids and bases and three independent reaction mechanisms account for neutral, acid, and base hydrolysis. Therefore, the overall hydrolysis kinetics has three contributing components.

$$\text{Rate of hydrolysis} = kh[\text{RX}] \tag{5.2}$$

$$\text{where } kh = kA[\text{H}^+] + kN + kB[\text{OH}^-], \tag{5.3}$$

the suffix A, N, and B with k refers to acid, neutral, and base hydrolysis.

Mechanism: Hydrolysis reactions of organic substrates are omnipresent in the environment. Hydrolysis is an important degradation reaction in surface, ground,

© Springer Nature Singapore Pte Ltd. 2022
K. Sudalyandi and R. Jeyakumar, *Biofuel Production Using Anaerobic Digestion*,
Green Energy and Technology, https://doi.org/10.1007/978-981-19-3743-9_5

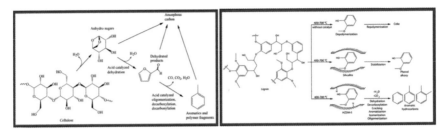

Fig. 5.1   a Reaction mechanism of cellulose [1]. **b** Lignin pyrolysis [2]

fog, and pore-water and can be a dominant pathway in biological systems as well. In general, hydrolysis occurs via one of two classes of mechanisms, based on the attachment of leaving group in the hydrolysis process:

(i) *Nucleophilic Substitution* (SN1 and SN2) generally occurs when the leaving group is attached to *sp3* hybridized carbon centre, such as alkyl halides, epoxides, and phosphate esters and

(ii) *Addition–Elimination*, generally occurs when the leaving group is attached to *sp2* hybridized acyl carbon centre, such as with carboxylic acid derivatives including esters, anhydrides, amides, carbamates, and ureas. Figure 5.1 shows the reaction mechanism of Cellulise and Lignin pyrolysis.

With the general information on kinetics and the mechanisms involved in the process of hydrolysis, the following sections detail various models that are available to describe the kinetics of hydrolysis.

5.1   Linear Regression

In this first-order hydrolysis kinetic relation, the hydrolysis rate is believed to be linearly proportional (at constant pH and temperature) to the amount of biodegradable substrate in the digester [3] as shown in Fig. 5.2.

$$-\frac{dC}{dt} = k\,C \tag{5.4}$$

k first-order hydrolysis constant;
C biodegradable substrate;
t time.

The mathematical description of the hydrolysis rate remains simple because the first-order hydrolysis constant combines all physical and enzymatic elements of the hydrolysis process. However, such an empirical connection does not help to a better knowledge or improvement of the actual hydrolysis process. For the hydrolysis

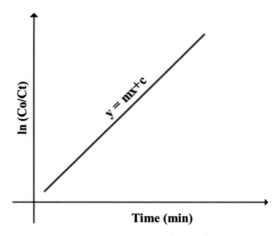

Fig. 5.2 First-order linear kinetic model

process, a linear first-order model is mostly used based on decomposable organic components.

5.1.1 Nonlinear Regression

Biodegradability and the rate of hydrolysis may be determined using nonlinear regression using a first-order kinetic model. An empirical first-order relation is a most basic and widely used perfect relationship [4].

$$M_f = M_{fd}\left(1 - e_{hyd}^{(-kt)}\right) \tag{5.5}$$

where M_{fd} is CH$_4$ produced as a result of organic decomposition (the portion of the organic substances which can be transformed to CH$_4$), k_{hyd} is degradation constant (day^{-1}), M_f is the yield of methane at the required period of degradation, and t is the period of degradation (days). Implementation of model in MATLAB 2012a, as well as the corresponding coding in analysis and predicting. In order to conduct a precision inquiry and a factor improbability analysis, the parameter assessment and implausibility assessment were calculated with a 95% confidence region.

5.2 Michaelis–Menten Model

As shown in Eq. (5.3), Michaelis–Menten kinetics, a well-known model of enzyme kinetics, is also used to describe the rate of hydrolysis based on the enzyme and substrate concentrations [5].

$$\frac{dS}{dt} = V_{max}\left(\frac{S}{Km + S}\right) \text{ with } V_{max} \qquad (5.6)$$

where S is the concentration of dissolved substrate concentration (g/L);

S is the enzyme concentration (g/L);

V_{max} is the maximum conversion rate (g/L/d);

K_m is the half velocity constant (g/L).

This model can be used to calculate two important parameters: V_{max} and K_m. V_{max} is the maximum product formation rate of substrate concentration at the saturation point which measures the catalytic potential of the enzyme. The Michaelis constant (K_m) is the substrate concentration at which the rate of reaction is half that of V_{max}, and it is commonly used to assess the active site's affinity for the substrate (the K_m value is inversely proportional to the affinity). In general, K_m and V_{max} are calculated by measuring an enzyme's initial reaction rate at various substrate concentrations. The reaction rate is then plotted against the concentration to form a Michaelis–Menten plot. By reciprocating both axes on the Michaelis–Menten plot, the V_{max} and K_m can be calculated from Lineweaver–Burk plot's line of best fit which is shown in Fig. 5.3.

This technique is based on mass conservation theories, which are largely applied to observed values reactions conditions, and so might have contributed significantly to the heterogeneity reaction mixture of enzymatic hydrolysis of resistant cellulosic substrates. The amount of absorbed cellulase is proportional to the concentration of substrate, and dual exhaustion may be achieved by keeping the enzyme or substrate concentrations high; these are not typical Michaelis–Menten kinetic features.

A heterogeneous cellulose hydrolysis process takes place on the substrate surface, resulting in proportions smaller than three. The standard chemical kinetics assumption of consistently mixed systems may not always apply to heterogeneous reaction systems, resulting in apparent rate orders, time-dependent rate constants, and non-uniform concentrations. In a cyclic or dimensionally limited media, changes of

Fig. 5.3 Michaelis–Menten model

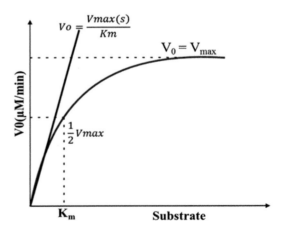

reactants or products. This sort of behaviour is referred to as fractal kinetics. The quasi-steady-state assumption cannot be utilized in these reaction systems, according to the Monte Carlo simulations. The conversion of cellobiose to glucose by glucosidase can be simulated using Michaelis–Menten kinetics since it is a homogenous process.

5.3 Stochastic Hydrolysis Model

Another approach for cellulose hydrolysis capable of capturing variable enzyme–substrate intermingling in hydrolysis is stochastic molecular modelling (SMM), which converts each hydrolysis event into a distinct event. This approach, which depends on molecular and enzymatic modelling of the enzymatic hydrolysis process, has been effective in explaining starch hydrolysis. This stochastic molecular modelling may be utilized to forecast hydrolyzed patterns and simultaneously resolve kinetics model constraints (For example, mathematical difficulties, a large number of parameter estimations, parameter changes when enzyme or hydrolysis conditions vary, and so forth) [6]. One of the main advantages of the SMM approach is that structural and enzyme features may be evaluated separately and modelled. The following are some of the additional advantages of SMM models.

- Modifications in substrate qualities and enzymatic attributes could be implemented without such requirement for additional testing.
- A large number of oligomer concentrations may be recorded significantly increasing the model's intricacy.
- Throughout hydrolysis, alterations in system parameters such as the proportion of chain ends, overall interfacial adhesion, and crystalline phase could be followed throughout time using the SMM approach, providing for a good sense of the efficacies and behaviours of enzymes.

To use this approach, the model must have a thorough and exact description of substrate features and enzyme function. Although the SMM technique needs more calculation, it is much more reliable and therefore can quickly compensate for discrepancies (substrate, enzyme, or reaction conditions).

The only first model for hydrolysis of refractory polysaccharides was conjured up using a stochastic modelling method. It was a constrained SMM model because it may not account for cellulose's actual structures (such as crystallinity and fibril structure) as well as the dynamics of several enzymes. Only a two-dimensional lattice depicting a specific cellulose area with a chain length of 20 was used in the hydrolysis. Since it was a deductive approach, the outcomes weren't validated by observational evidence. The previously used similar technique to model cellulose hydrolysis was also possible to forecast endoglucanase and CBH enzyme hydrolysis patterns with high precision [7]. To that extent, the approach is capable of capturing dynamic enzyme–substrate interactions. The model, unfortunately, excluded essential cellulose structural features and also had limited scope. It was a two-dimensional model in which those enzymes could access all glucose chains, that is not the case in this instance. Since a crystalline structure mostly in cellulose and hemicelluloses might not have been taken into consideration the activity difference among both enzymes in crystalline and amorphous zones. The model did not account for inhibition by cellobiose or glucose, which would be a significant parameter throughout cellulose breakdown.

5.4 Simplified HCH-1 Model

The basic Holtzapple–Caram–Humphrey-1 (HCH-1) model for the enzymatic hydrolysis of cellulose (C) was revised by Liang et al. [45] to create a long-term HCH-1 model for the enzymatic hydrolysis of cellulose (C). As hydrolysis products alter reaction rates, the earlier technique only evaluated quick reaction rates, making it meaningless for long-term forecasts. This influence on cellulosic conversion efficiency and enzymatic durability during extended time periods is compensated for the modified HCH-1. The usage of the product binding constant and the assessment of the number of active sites covered by enzymes (E) using the parameter are two of its key aspects. The cellulose conversion factor is x where the HCH-1 model is totally dependent on adsorption mechanics since it incorporates all of these adsorption processes and Michaelis–Menten kinetics. In fact, towards inhibiting the activity anticipated in the model, it incorporates quasi-restriction of enzymatic activity through the hydrolysis products formed. The modified HCH-1 model varies from the initial formulation because it includes aggregated enzymatic mixtures rendering it much more authentic.

5.5 Semi-Mechanistic Model

Semi-mechanistic models feature a realistic adsorption model and are based on a single cellulose hydrolyzing activity and/or dependent on concentration as the only variable defining the state of the substrate. Semi-mechanistic models with concentration as the only substrate variable such as crystallinity and considering the action of multiple enzymes in hydrolysis are said to be semi-mechanistic in terms of substrate, whereas semi-mechanistic models with a single cellulose hydrolyzing activity are said to be semi-mechanistic in terms of enzyme. The majority of hydrolysis models presented so far for industrial system design fall under the semi-mechanistic group. Semi-mechanistic models can be beneficial in the context of exercises that are driven by incorporating just the information required for descriptive purposes. The semi-mechanistic substrate models, on the other hand, never explain or provide the input into behaviour that is influenced by substrate characteristics other than concentration. Similarly, semi-mechanistic enzyme models could somehow describe or give insight into the actions regulated by diverse hydrolyzing processes.

Semi-mechanistic enzyme models only use factors other than substrate concentration to describe the condition of the substrate. Many models in this category appear to be motivated by the commonly observed tendency of decreasing rate with greater conversion. There are many models detailing an expected change in surface area and shape during hydrolysis. None of these models have been tested against experimental data as found in the literature (for example, surface area). Several "two-substrate" hypotheses have been proposed, which divide cellulose into a less reactive highly crystalline component and a more reactive amorphous proportion. Even while such models have shown some promise in terms of data correlation, the relationship between increasing CrI and higher conversion, as expected if amorphous cellulose responds first, has yet to be substantiated by experimental evidence [8]. A two-substrate model includes shrinking cellulose spheres with only an amorphous shell and a shrinking core, and also cellobiose cellulose hydrolysis inhibition and h-glucosidase cellobiose liquid-phase hydrolysis inhibition.

5.6 Non-mechanistic Approaches

The non-mechanistic models provide correlations for fractional conversion or response rate as a function and numerous numbers of parameters. Models with conversion as the outcome include parameters such as enzyme loading and substrate concentration, as well as pretreatment biomass characteristics. Hydrolysis time, enzyme loading, and cellulose conversion are all factors that are integrated into models using rate as the output. Non-mechanistic models were established prior to the early 1980s. The model provided an example of one using conversion extent as an output. For example, the influence of pretreated wheat straw qualities on cellulose conversion using nonlinear multivariate regression to develop an exponential model (X = 1 − [final cellulose concentration]/[initial cellulose concentration]) after 8 h.

5.7 Hill and Barth Model

Hill and Barth formulated the model in 1977 in which the hydrolysis, acidogenesis, and ammonia inhibition process was considered. Figure 5.4. In 1982, Hill's model suggests that methanogenesis is dependent on total fatty acids. This model was created specifically to describe the digestion of manure and animal waste. The total fatty acid concentration is assumed to impede the model.

The bacterial groups listed below are assumed to be involved in the entire digestion process (shown in Fig. 5.5): (a) Acidogenic bacteria that feed on glucose (dissolved organics minus volatile fatty acids) and create a combination of acetic, propionic, and butyric acids; (b) hydrogenogenic organisms that develop slowly and convert propionic and butyric acid into acetic acid and H_2; (c) homoacetogenic produces

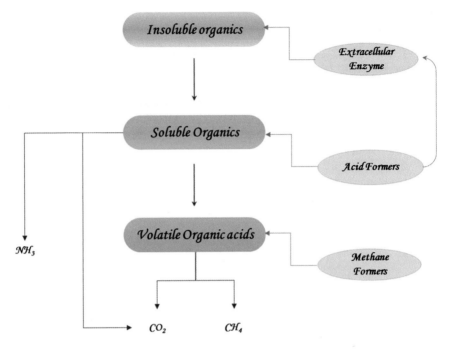

Fig. 5.4 Hill and Barth model

acetate from H_2 and CO_2; (d) H_2-methanogenic converts CO_2 into CH_4; and (e) (CH_4 and CO_2).

High fatty acid concentrations are thought to hinder all five processes. This inhibition may be seen in both the growth rate and the microbial degradation rate. When a buildup of VFAs occurs, anaerobic digestion is halted, according to this concept. For instance, inhibition reduces the rate of VFA consumption, resulting in acid buildup. The digester fails regardless of pH value above a specific threshold VFA content. For each of the five primary reactions, this model is based on stoichiometric reactions. Because the majority of stoichiometric coefficients and numerous kinetic rates were not found and these parameters can be determined by fitting pilot-scale and full-scale anaerobic digesters.

5.8 Simplified Kinetic Model [9, 10]

(i) Assuming the reaction is occurring at the liquid–liquid interface and the relevant kinetics equation can be formed as detailed below:

$$(1)\ T + CH_3O^- \rightarrow D^- + E \quad r_1' = k_1 a \left[CH_3O^- \right]_p [T]_a \qquad (5.7)$$

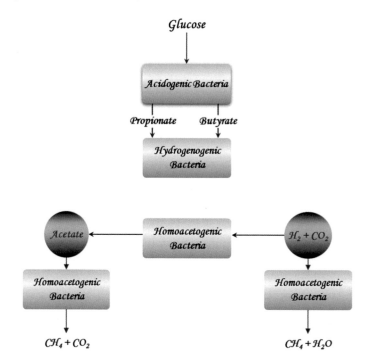

Fig. 5.5 Modified Hill model

$$(2) \quad D + CH_3O^- \rightarrow M^- + E \quad r_1'' = k_1 a [CH_3O^-]_p [D]_a \qquad (5.8)$$

$$(3) \quad M + CH_3O^- \rightarrow G^- + E \quad r_1''' = k_1 a [CH_3O^-]_p [M]_a \qquad (5.9)$$

p and **a** are polar and *apolar* phase, respectively. The kinetic constants for the three reactions occurring at the interface have been considered to be equal.

(ii) Assuming reactions occurring in *apolar* phase

$$(4) \quad D^- + CH_3OH \rightarrow M^- + E \quad r_2 = k_2 [D^-]_a [CH_3OH]_a \qquad (5.10)$$

$$(5) \quad M^- + CH_3OH \rightarrow G^- + E \quad r_3 = k_2 [M^-]_a [CH_3OH]_a \qquad (5.11)$$

$$(6) \quad T + CH_3OH \rightarrow D + E \left(\text{Catalyst} \left(D^- + M' \right) \right)$$
$$r_4 = k_4 [D^- + M^-]_a [T]_a [CH_3OH]_a \qquad (5.12)$$

$$(7) \quad D + CH_3OH \rightarrow M + E \left(\text{Catalyst} \left(D^- + M' \right) \right)$$

$$r_5 = k_5[D^- + M^-]_a[D]_a[CH_3OH]_a \tag{5.13}$$

(8) $M + CH_3OH \rightarrow G + E\left(Catalyst \left(D^- + M'\right)\right)$

$$r_6 = k_6[D^- + M^-]_a[M]_a[CH_3OH] \tag{5.14}$$

Reactions 5.6, 5.7, and 5.8 have been written neglecting the intermediate forma-tion of the *enolate* species by suppressing that their concentrations are low as their disappearing velocity is very fast.

(iii)   Reaction occurring in polar phase:

(9) $CH_3OH + G \rightleftarrows CH_3O^- + G$ $r_7 = k_7\left\{[CH_3OH]_p - \dfrac{[CH_3O^-]_p[G]_p}{k_{eq,7}}\right\}$

$$\tag{5.15}$$

The dependence of the kinetic constants on the temperature can be expressed in all cases by the relationship

$$k_i = k_i^{ref} exp[\frac{E_{ai}}{R}(\frac{1}{T^{ref}} \frac{1}{T})] \tag{5.16}$$

Reaction (5.9) could be considered near to the equilibrium. Methanol and Glycerol are more intensively transferred between the two phases with respect to the other compounds, and the equations describing the mass transfer rates can be formulated as follows:

Polar phase

$$J_{CH_3OH}^P = k_{l,CH_3OH} \cdot a \cdot ([CH_3OH]^P - H_{CH_3OH} \cdot [CH_3OH]^{a,*}) \tag{5.17}$$

$$J_G^P = -k_{l,G} \cdot a \cdot ([G]^P - H_G \cdot [G]^{a,*}) \tag{5.18}$$

$$J_{G^-}^P = -k_{l,G^-} \cdot a \cdot ([G^-]^P - H_G \cdot [G^-]^{a,*}) \tag{5.19}$$

Apolar **phase**

$$J_{CH_3OH}^a = k_{l,CH_3OH} \cdot a \cdot ([CH_3OH]^{a,*} - [CH_3OH]^a) \tag{5.20}$$

$$J_G^a = -k_{l,G} \cdot a \cdot ([G]^{a,*} - [G]^a) \tag{5.21}$$

$$J_{G^-}^a = -k_{l,G^-} \cdot a \cdot ([G^-]^{a,*} - [G^-]^a) \tag{5.22}$$

The interfacial concentration (denoted by suffix *) can be calculated by solving the following equations that consider a steady-state condition, which is always valid for the component J

$$J_J^P \cdot V^P = J_J^a \cdot V^a \tag{5.23}$$

It can then be written as

$$[J]^{a,*} = \frac{V^P \cdot \beta j \cdot [J]^P + \beta j \cdot V^a \cdot [J]^a}{\beta j \cdot V^a + V^P \cdot \beta j \cdot HA} \tag{5.24}$$

$$[J]^{P,*} = H_J \cdot [J]^{a,*} \tag{5.25}$$

With $\beta j = k_{l,j} \, a$.

To solve the mass transfer equations, it is essential to define the partition parameters $H_J = [J]^a / [J]^P$, for both methanol and glycerol. The solubility parameters have been estimated by using ChemCAD V 6.3 version and UNIFAC LLE model, and the evolution of these parameters with triglycerides conversion (x) has been estimated and expressed with the two following polynomial equations:

$$\frac{1}{H_{CH_3OH}} = 25.85 - 23x - 11.21x^2 + 36.41x^3 \tag{5.26}$$

$$\frac{1}{H_G} = 6304.75 - 15430.10x - 1643.71x^2 + 64912.73x^3$$
$$- 92127.03x^4 + 43103.45x^5 \tag{5.27}$$

5.9 Simple Method of Determining the Kinetics Parameters of Biphasic Reactions

The hydrolysis of biomass is often considered by researchers as Pseudo-first-order reaction, and the speed of the hydrolysis process is fast instantaneously at the beginning and becomes slower later. These two different phases of reaction couldn't be described well by other models but the biphasic reaction model described well the process and accurately fit the experimental data [11, 12].

This section describes a simple method of determining the rate constant of biphasic reaction. In an example of the hydrolysis process of biomass, first draw a graph of % residual activity versus time. Figure 5.6 shows two distinct phases of reaction, i.e. fast (A) and slow (B). In the slow phase, one can observe a slow exponential phase, and in fast (A) phase, the graph rapidly drops downward linearly with time. Drawing

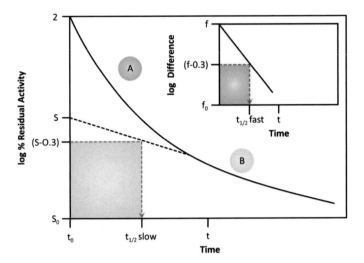

Fig. 5.6 Semilog plot for the determination of the rate constants of fast (A) and slow (B) phases of a biphasic reaction

Tangent from the slow phase curve that meets at zero time and correction for the same reveals that the fast phase follows first-order kinetics as shown in (insert) Fig. 5.6.

The complete reaction process is explained by the following expression:

$$C_t = C_{\text{fast}} \cdot e_{\text{fast}}^{-k.t} + C_{\text{slow}} \cdot e^{-k_{slow}.t} \tag{5.28}$$

where C_t = the residual activity at time t;

C_{fast} and C_{slow} = Amplitudes as observed from the graph;

k_{fast} and k_{slow} = First-order rate constants.

The tangent drawn from the slow phase curve meets at Y-axis at point S, and the antilog value of that S gives the value of the slow phase.

$$\text{Then} \quad \%C_{\text{fast}} + \%C_{\text{slow}} = 100\% \tag{5.29}$$

Percentage C could be estimated from Eq. 5.2 and the half-life of the slow phase can be estimated from Y-axis at S-0.3 and the corresponding value against the X-axis is $t_{\frac{1}{2} \text{ slow}}$ as shown in Fig. 5.6. The kinetic rate is determined from the following equation:

$$k = 0.69/t_{1/2} \tag{5.30}$$

By plotting the difference between observed and extrapolated values of the residual activity of the fast phase versus time as in Fig. 5.6 (inset), kfast can be determined. In the figure, the straight line starts at the point f. Draw a horizontal at (f-0.3)from the y-axis to the reaction line and from the point of intersection draw a vertical to the x-axis (time). This value of time is the half time of the fast phase ($t_{1/2 \text{ fast}}$) and by applying Eq. 5.3 k_{fast} can be determined.

References

1. Carlson TR, Tompsett GA, Conner WC, Huber GW (2009) Aromatic production from catalytic fast pyrolysis of biomass-derived feedstocks. Top Catal 52(3):241–252. https://doi.org/10.1007/s11244-008-9160-6
2. Ma Z, Troussard E, Bokhoven JAV (2012) Controlling the selectivity to chemicals from lignin via catalytic fast pyrolysis. Appl Catal A 423–424:130–136. https://doi.org/10.1016/j.apcata.2012.02.027
3. Wang J, Ye J, Yin H, Feng E, Wang L (2012) Sensitivity analysis and identification of kinetic parameters in batch fermentation of glycerol. J Comput Appl Math 236:2268–2276
4. Eastman JA, Ferguson JF (1981) Solubilization of particulate organic carbon during the acid phase of anaerobic digestion. J Water Pollut Control Fed 53:352–366
5. Pletschke BI, Rose PD, Whiteley CG (2002) The enzymology of sludge solubilisation utilising sulphate reducing systems. Enzyme Microb Technol 31:329–336
6. Yu L, Wensel PC (2013) Mathematical modeling in anaerobic digestion (AD). J Bioremediat Biodegrad 4:1–12
7. Neto JM, Dos Reis GD, Rueda SMG, da Costa AC (2013) Study of kinetic parameters in a mechanistic model for enzymatic hydrolysis of sugarcane bagasse subjected to different pretreatments. Bioprocess Biosyst Eng 36:1579–1590
8. Saha BC, Iten LB, Cotta MA, Wu YV (2005) Dilute acid pretreatment, enzymatic saccharification and fermentation of wheat straw to ethanol. Process Biochem 40:3693–3700
9. Santacesaria E, Russo V, Tesser R, Di Serio M (2019) A Kinetic Biphasic approach to biodiesel process intensification. Chem Eng Trans 74:1339–1344
10. Dijkstra AJ, Toke ER, Kolonits P, Recseg K, Kovári K, Poppe L (2005) The base-catalyzed, low temperature interesterification mechanism revisited. Eur J Lipid Sci Technol 107:912–921
11. Ray WJ, Koshiland DE (1961) Brookhaven Symp Biol 13(139)
12. Kayastha AM, Gupta AK (1987) An easy method to determine the kinetic parameters of biphasic reaction. Biochem Educ 15(3)

Chapter 6
Fermentation

Fermentation refers to the foaming that occurs during the production of wine and beer, a process that has been occurring for at least 10,000 years. The development of carbon dioxide gas, which was not known until the seventeenth century causes foaming in fermentation. Louis Pasteur, a French chemist and microbiologist, coined the term fermentation to describe the difference between yeasts and other microorganisms that grow in the absence of oxygen (anaerobically) in the nineteenth century. In the mid-1850s, a French chemist named Louis Pasteur invented anaerobiosis by boiling the fermentation medium to eliminate oxygen. He demonstrated that butyric acid fermentation was caused by a microbe, most likely *Clostridium butyricum*. The anaerobic generation of ethanol by *S. cerevisiae* and other yeasts is the most important industrial fermentation. During fermentation, the anaerobes contain typical enzymes and catabolic pathways to limit contamination during fermentation.

6.1 Theory of Fermentation

Carbohydrates (for example glucose) are used by microbes for the synthesis of energy and fuel. During fermentation, the required energy for the process was supplied by ATP (adenosine triphosphate) and this energy was provided to all parts of the cell. Bacteria generate ATP via respiration. The most efficient method is aerobic respiration, which requires oxygen. Aerobic respiration begins with glycolysis, which converts glucose to pyruvic acid [1]. Other organic acids such as lactic acid are also formed during fermentation that produces ATP. Fermentation can take place in the absence of oxygen (anaerobic conditions) and in the presence of microorganisms (yeast, mould, and bacteria) that get energy from the fermentation process. These microbes convert sugars and starches into alcohols and acids during fermentation.

During fermentation, through the metabolism of glucose ($C_6H_{12}O_6$), often known as starch or glycolysis, yeast converts pyruvic acid molecules into molecules of alcohol and carbon dioxide. The pyruvic acid uses nicotinamide adenine dinucleotide

© Springer Nature Singapore Pte Ltd. 2022
K. Sudalyandi and R. Jeyakumar, *Biofuel Production Using Anaerobic Digestion*,
Green Energy and Technology, https://doi.org/10.1007/978-981-19-3743-9_6

+ hydrogen (NADH) to generate NAD $^+$ with lactic acid in an anaerobic biological process. Bacteria generate ATP by breaking down one molecule of glucose for every two molecules of pyruvic acid. The major products produced during fermentation are acetic acid (CH_3COOH), hydrogen (H_2), and carbon dioxide (acetic acid synthesis) (CO_2).

6.2 Factors Affecting Acetogenesis

6.2.1 PH Control

Fluctuation of pH during acetogenesis is mediated by fermentative metabolites such as VFAs, ammoniacal nitrogen, and bicarbonates. pH is often an essential element in AD as it alters the biochemical route and the product generated. In an AD system, homoacetogenic bacterial population grows well in unfavourable pH range of 5.9–6.6. An acidic pH, on the other hand, is beneficial to the acetogenesis. When the acidity of the water rises, the carbonate is converted to carbon dioxide. As a result, a low-carbon and low buffering capacity state are formed. The occurrence of an acidic state during fermentation is considered to be a better option to favour a certain organic acid. At pH 5, acetate appeared to be a rather prevalent product, preceded by butyrate and propionate. Additionally, an alkaline pH of 7.2–8.0 was more suitable for CH_4 production during hydrogenotrophic methanogenesis.

6.2.2 Temperature

The operating temperature of the AD system is a crucial parameter that determines the growth and metabolic activity of the microbes. The temperature has a three-dimensional impact on the AD system since it influences microbial growth rate, complicated substrate hydrolysis, and enzymatic action. The AD process can be three types based on temperature, thermophiles above 50 °C, mesophiles between 30–45 °C, and psychrophiles below 20 °C. In the case of a complex substrate, the temperature can be used to choose a preferable metabolic route for H_2 substrate absorption [2]. Lesser operational temperature favours the growth of H_2 methanogens. Sulphate-reducing bacteria (SRB) can outperform hydrogenotrophic methanogens at a mesophilic temperature of 37 °C; but, at a thermophilic temperature of 55 °C, hydrogenotrophic methanogens are the dominating microbial population. The thermophilic temperature (65 °C) boosted CH_4 generation by acetate oxidation and hydrogenotrophic methanogenesis. The thermophilic temperature has been proven to be beneficial in enhancing the microbial growth and reducing the retention time of the system. The two demerits of thermophilic temperatures in the AD system are an increase in bacterial load and a lack of microbial diversity.

6.2.3 Microbial Cultures

The microbial cultures (inoculum) are an important parameter to be determined in the AD system. An inoculant is a culture that contains fermentative microorganisms. It is divided into two types: mixed inoculum and pure inoculum. In the case of simple substrates, specific microbes can be used as inoculum. For example, yeast (mainly *Saccharomyces cerevisiae, Pichia stipites, Kluyveromyces fagilis*, and *Zymomonas mobilis*) is commonly used in ethanol fermentation and *Lactobacillus* is commonly used in lactic acid fermentation [3]. Yeast is a facultative single-celled fungus that can grow in both aerobic and anaerobic conditions. When there is a lack of oxygen, yeast generates energy by utilizing carbohydrates and converting them into CO_2 and ethanol.

In fermentation, *S. cerevisiae* grows well in the pH range of 4.0–5.0 and at a temperature of 30 °C. The Gram-negative facultative anaerobe *Zymomonas mobilis* thrives at temperatures ranging from 25 to 31 °C. The pH level ranging from 3.5 to 7.5 is favourable for *Zymomonas mobilis* and seems to have high acid resistance. This yeast does not contain the enzyme 6-phosphofructokinase and it uses the Entner–Doudoro (ED) route to metabolize glucose, fructose, and sucrose, producing ethanol and CO_2. This yeast does not use glycolytic pathways. The ED pathway generates 50% less ATP and it generates 12 percent (w/v) more ethanol at a faster rate than the Embden–Meyerhof–Parnas (EMP) route.

Lactobacilli are Gram-positive, anaerobic or facultatively anaerobic non-bacilli, and these bacteria grow in the temperatures ranging from 20 to 45 °C and pH levels ranging from 5 to 7. They are effective fermentative microbes. *Clostridium sps* are preferred for the synthesis of butyric acid or butanol. These bacteria are strictly anaerobic, gram-positive, and chemo-organotrophs and grow in temperatures of 35–37 °C and pH of 4.5–7.0.

Usage of pure culture is difficult to treat complex waste. Because the existence of indigenous bacteria and maintaining aseptic conditions are both expensive on a large scale. A mixed culture has greater microbial variety than a pure culture, requires less ambient conditions, which is more resistant to environmental variations, and can operate continuously without strain deterioration. To enrich certain microorganisms and increase the synthesis of target compounds in mixed inoculation, certain changes in environmental conditions and extra inoculant addition (such as yeast) have to be used. Mixed culture has more microbial variety than pure culture, requires less adjustment in environmental parameters and is more robust to environmental fluctuations, and may be operated indefinitely without strain degeneration. The mixed inoculum comprises a high concentration of facultative anaerobes, which might help the organism to sustain in anaerobic conditions in the aspect of high redox potential. In the case of mixed cultures, the presence of acetoclastic methanogen can help convert surplus acetate to methane when it accumulates in a hydrogenotrophic dominant reactor. Although it is necessary to introduce an H_2 substrate absorption process with a large inoculum supply, this does not ensure the system's long-term stability.

6.3 Strategies to Enhance Acetogenesis

6.3.1 Addition of Enhancing Agents (Biosurfactants, Zerovalent Iron (ZVI), Etc.)

The most typically employed additives are iron-based compounds and carbon-based elements. The former includes ZVI (Zero valent ions), iron oxide, and magnetite (including Nano forms), while the latter includes activated carbon and biochar. ZVI produces a significant number of micro-electrolysis processes in fermenters, which results in rapid dissolution and hydrolysis of particulate organic matter. ZVI may reduce (oxidative reduction potential) ORP, allowing facultative or obligatory fermentative bacteria to thrive in a low-ORP environment [4]. The ZVI can also boost the pH of fermentation systems by interacting with H^+, which improves enzyme activity. The liberated ions in the AD system have the potential to increase the activity of pyruvate-ferredoxin oxidoreductase, an enzyme that is involved in the oxidative decarboxylation of pyruvate to acetyl-CoA during the production of acetic and butyric acids in the fermentation process.

During anaerobic metabolism mediated by dissimilatory iron-reducing bacteria, iron oxides (Fe_2O_3, Fe_3O_4) frequently serve as the final recipient of electrons, allowing the breakdown of complex organic compounds and related enzyme activity with liberated ferrous ions. The presence of Fe (III)-reducing species such as *Clostridium species* can aid FW hydrolysis.

Activated carbon is a typical carrier for microbial immobilization due to its high surface area, great mechanical characteristics, low toxicity, and inertness. The activities of protease, dextranase, and lipase were boosted by activated carbon at a dose of 0.5 g/L, and the quantity of Proteobacteria, which can break down organic molecules into acetic acid and other products, increases considerably. Biochar is a low-cost carbonaceous material with a high specific surface area and microporosity [5]. Biochar can assist the growth of fermentative bacteria by not only acting as a substrate for microbial adhesion and biomass development but also by releasing trace metals and volatile compounds. The addition of 1.0 g/L of charcoal decreases the bacterial diversity, favouring *Firmicutes* and *Bacteroidetes* that produce anaerobic VFA while suppressing *Proteobacteria*.

6.3.2 Reactors

Continuous stirred tank reactors (CSTRs) are frequently employed due to their simple design, simplicity of modifying working parameters, and stirring, which guarantees that the medium is homogenous and microorganisms have good contact with substrates. Stirring also enables for exact control of parameters inside the reactor (pH and temperature). The yield of hydrogen was increased with the decrement in the partial pressure which happens due to the continuous stirring of the reaction mixture to enhance the process efficiency. Bacterial existence in CSTR has better suspension in liquid substrate than in other fermenters. In this CSTR, the major critical factors governing the process are the HRT and organic loading rate. Retention time highly depends on HRT which restricts the biomass concentration in the reaction mixtures and has an obvious effect on hydrogen production rates [6]. In case of high dilution, biomass removal emerges as a problem, and microbes lose their capacity to function. The growth rate of biomass was found to be maximum at higher HRT. For shorter HRT, the hydrogen generation rate within the CSTR reaction will be lower.

A membrane bioreactor (MBR) is a type of reactor that has a membrane or a membrane system linked to it and is used to carry out fermentation activities. In this MBR, the proper positioning of the membrane was done using cross-flow which was placed outside the stream of the reactors or submerged. Since it has lower operational costs and smaller membranes, the latter type is preferred. The use of a membrane bioreactor overcomes the CSTR's principal drawback: biomass elution. The membrane keeps microorganisms contained within the reactor, allowing for a consistent, high biomass concentration. The membrane allows for the selection of an optimum HRT that is independent of the sludge retention duration, allowing for better control of process parameters [7]. Increased retention time boosts biomass HRT, which boosts substrate conversion but lowers hydrogen generation. Membrane fouling and high depletion of cost on operating the side-stream membrane systems are two key disadvantages of membrane bioreactors.

CSTRs can be replaced by packed-bed reactors. Unlike CSTRs, it may be used to convert biomass that contains a significant amount of organics. Shorter HRT with nil biomass elution characterizes the performance of the reactor. The bed of the reactors is made of biofilm granules or gel (microbes bind to form a gel-like network). Packed-bed reactors include anaerobic fluidized-bed reactors (AFBR), upflow anaerobic sludge blanket (UASB) reactors, and expanded granular sludge bed (EGSB) reactors. Immobilization of microbes on the surface of the packed bed improves the fermentation process. The microbes are immobilized on the materials which are spongy in nature having 552 mm diameter and on the glass material of diameter 7 mm within the continuous packed column to enhance the dark fermentation [8]. This immobilization limits the wastage of microbial biomass from the reactor and minimize the microbial stress developed during stirring process.

The combined CSTR and immobilized bed reactors consist of deposited granules as a biofilm. The advantage of AFBR is it assists in the transmission of heat and mass which deliver gas that enables the formation of a fluidized bed from the reactor bottom. This gas formation hinders the catalytic activity affecting the degradation of the substrate since the deposited microbes may get washed away in the UASB reactor. An AFBR's disadvantage is the increased energy consumption which is necessary to maintain the bed fluidized. The three-phase separator installed to the upper part of the bioreactor is the major component of the system. As an influent distributor, a bed of 5 mm glass beads was placed at the reactor's bottom. As a support medium, the bacteria were cultivated on granular activated carbon. A heating blanket was used to keep the pH at 4.0 and the temperature of the reactor at 37 °C [9].

The UASB reactor is typically elongated and has three-phase separator in the upper portion at which hydrogen-producing granules (HPG) are rapidly generated and deposited towards the bottom forming a thick layer of biomass. The activated sludge aggregates form granules that are produced throughout the fermentation process. Its diameters range from 0.2 to 2 mm. Granulated microorganisms have a higher retention rate in the reactor and are more resistant to unfavourable environmental conditions. No mechanical stirring is required since the formation of gas bubbles during fermentation ascended which causes turbulence in the medium when the feedstock is pumped to the bottom of the reactor. The major disadvantage of UASB in using granulated activated sludge is a substrate takes a long start time. UASB reactor is sensitive to channelling impact that results in the least interactions between substrate–biomass. These reactors are capable of processing a wide range of substrates, including organic waste from a variety of sources. This type of reactor generates hydrogen with excellent efficiency and consistency. UASB reactors have a high efficiency of hydrogen generation with minimum HRT and constant operational conditions. In addition, mostly in two-stage systems, hydrolysis and fermentation take place independently. Fermentation can also be carried out in a succession of reactors [10, 11]. This approach provides more feedstock conversion and process yield, but at the expense of a longer total retention time and higher system running costs.

6.3.3 Environmental Parameters

Fermentative bacteria are less sensitive to pH and may function in a wider pH range between 4.0 and 8.5: acetic and butyric acid are the primary products at low pH, while acetic and propionic acid are the predominant products at a pH of 8.0. The pH is lowered by the VFAs generated. Methanogenic bacteria, which create alkalinity in the form of carbon dioxide, ammonia, and bicarbonate, help to offset this loss. The pH of the system is determined by the CO_2 concentration in the gas phase and the HCO^{3-} alkalinity in the liquid phase. Adding HCO^{3-} alkalinity to the digester can raise the pH if the CO_2 content in the gas phase remains constant. For a stable and well buffered digestive process, a buffering capacity of 70 meq $CaCO_3$/l or a bicarbonate/VFA molar ratio of at least 1.4:1 should be maintained, even though it has been proven that the ratio's stability, rather than its level, is more important. The physicochemical qualities of the components contained in the digesting substrate are influenced by temperature [12]. In the anaerobic reactor, it also impacts the rate of microbe growth and metabolism, as well as population dynamics.

The temperature increase is particularly harmful to autotrophic methanogens. Propionate and butyrate breakdown are both harmed by temperatures above 70 °C. Furthermore, temperature influences the partial pressure of H_2 in digesters, which affects syntrophic metabolic kinetics. The breakdown of propionate into acetate, CO_2, and H_2 becomes more energetically advantageous at higher temperatures, according to thermodynamics, whereas exergonic processes (such as hydrogenotrophic methanogenesis) become less favourable [13].

6.4 Kinetics and Modelling of Acetogenesis

6.4.1 Andrews Model

To determine the performance of the anaerobic reactor, several mathematical models have been constructed. Andrews in 1969 discovered a dynamic model to simulate a digester's startup and failure [13–16]. In the AD process, the conversion of fatty acids into biogas is thought to be a bottleneck. Through this model, Andrews proposed that the unionized acids act as both growth limiting and inhibitory agent. Graef and Andrews introduced an inhibition function in 1973 to determine the inhibition of AD. In this model, the composition of methanogens is assumed to be $C_5H_7NO_2$ and volatile fatty acids are given as acetic acid. According to this model, the overall reaction can be represented as follows:

$$CH_3COOH + 0.032\,NH_3 \rightarrow 0.032\,C_5H_7NO_2 + 0.92\,CO_2 + 0.92\,CH_4 + 0.096\,H_2O \quad (6.1)$$

6.4.2 Mosey Model

Mosey (1983) identified the hydrogen partial pressure as a critical regulator of anaerobic biodegradation of glucose which was shown in Fig. 6.1. This has an impact on the redox potential in the liquid phase of the substrate in AD. In this model, the function of four bacterial groups in the conversion of glucose to CO_2 and CH_4 was proposed: (a) Acid-forming bacteria that grow quickly ferment glucose to generate a combination of acetate, propionate, and butyrate; (b) Propionate and butyrate are converted to acetate by acetogenic bacteria; (c) Acetate is converted to CO_2 and CH_4 by acetoclastic methane bacteria; and (d) CO_2 is converted to CH_4 by bacteria that need hydrogen [17]. The relative production of fatty acid is hypothesized to be influenced by the redox potential, or more precisely, the ratio of [NADH]/[NAD$^+$].

This ratio is determined by the partial pressure of hydrogen in the gas phase. Since acidogenic bacteria follow the glycolytic metabolic route, the liquid-phase redox potential, or equivalently the ratio [NADH]/[NAD$^+$] inside the bacterial mass, determines the relative quantities of fatty acid synthesis. This ratio may be stated as a function of hydrogen partial pressure using the following assumptions:

1. Reflecting variations in the liquid phase, the bacteria retain a neutral pH within.

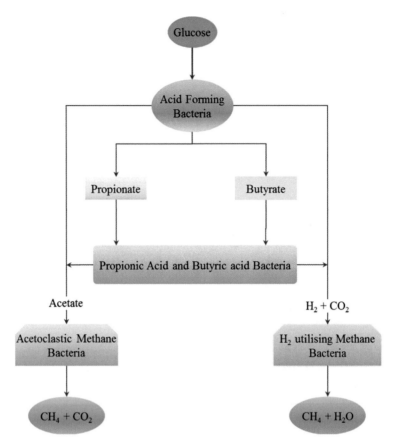

Fig. 6.1 Mosey model

2. Hydrogen gas diffuses readily and quickly across the bacterial membrane, resulting in a partial pressure inside the cell that is identical to that in the digester gas phase.
3. The interior of the cell has the same redox potential as the liquid media.

Besides acidogenic bacteria, the existence of high hydrogen partial pressure inhibits the synthesis of propionic and butyric acids, which slows acetogenic development (thermodynamically). Finally, low pH levels are likely to suppress all bacterial species (pH 6). According to the Mosey model, a rapid increase in the organic loading rate will result in a buildup of VFAs since acetogens develop at a slower rate than acidogens. The resulting pH drop inhibits the growth of methanogenic bacteria that utilize hydrogen producers, resulting in a rise in hydrogen partial pressure and more propionic and butyric acid buildup [18]. When the pH falls below 5.5, the methane production stops.

6.4.3 Bryers Model

The model is designed to study the hydrolysis of complex materials such as particulate organic matter into protein, carbs, and lipids and then into amino acids, simple sugars, and fatty acids. As a result, amino acids, carbohydrates, and fatty acids are produced in the first step, while intermediates such as propionate, butyrate, and acetic acid are produced in the second and third phases (acidogenesis) from amino acids/sugars and fatty acids. The fourth step is acetogenesis, whereas the fifth and sixth stages are methanogenesis that starts from acetate and hydrogen, respectively [16].

6.4.4 Angelidaki Model

This model explains all the processes involved in AD such as hydrolysis, acidogenesis, acetogenesis, and methanogenesis (Fig. 6.2). This model proposes that methanogenesis is inhibited by free ammonia, acetogenesis is inhibited by acetic acid, and acidogenesis is inhibited by total VFA [19].

Temperature and pH are affecting the bacteria's maximal specific growth rate and the degree of ammonia ionization. The mechanism for pH self-regulation is that the acetic acid accumulates whenever free ammonia (high for high pH) hinders methanogenesis. This inhibits acetogenesis, resulting in a buildup of propionic and butyric acids, and as a result, inhibition of acidification happens. The model determines the

Fig. 6.2 Angelidaki model

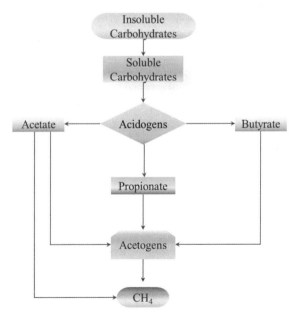

performance of digesters. VFA accumulation lowers the pH, which inhibits methano-genesis by lowering the free ammonia concentration [20]. When the rate of inhibition is greater than the system's tolerance, then the process can be self-regulatory. As a result, the pH decreases dramatically resulting in digester failure.

6.4.5 Siegriest Model

Siegrist created the Siegrist model shown in Fig. 6.3 which is a shortened version of the Siegrist model. To simulate the process, Siegrist employed municipal sludge, which is a combination of primary, secondary, and tertiary sludge collected from wastewater treatment plants [21]. The model assumptions are as follows:

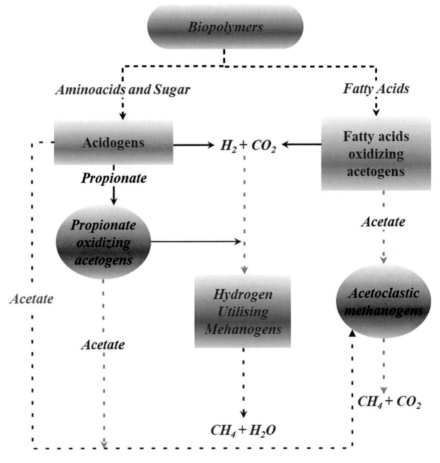

Fig. 6.3 Siegriest model

- As the temperature rises, the maintenance of the system increases and changes in stoichiometry also happen.
- The KLa value in mass transfer theory is solely dependent on temperature; however, it is modified by various factors.
- The digester is thought to be suspended totally, and the sludge retention period is the same as the hydraulic retention time.

Limitations

- The rapid temperature change cannot be calculated due to changes in biomass communities.
- Incorporating particle size distribution analysis into the model is a denial.

References

1. Saeman JF (1945) Kinetics of wood saccharification—hydrolysis of cellulose and decomposition of sugars in dilute acid at high temperature. Ind Eng Chem 37:43–52
2. Tao ZH, Wang SX, Ji LX, Zheng L, He W (2013) Electrochemical investigation of the adsorption behaviour of guanine on copper in acid medium. Adv Mat Res 787:30–34
3. Ulas S, Diwekar UM (2004) Thermodynamic uncertainties in batch processing and optimal control. Comput Chem Eng 28:2245–2258
4. Zhang Y-HP, Lynd LR (2004) Toward an aggregated understanding of enzymatic hydrolysis of cellulose: noncomplexed cellulase systems. Biotechnol Bioeng 88:797–824
5. Bousková A, Dohányos M, Schmidt JE, Angelidaki I (2005) Strategies for changing temperature from mesophilic to thermophilic conditions in anaerobic CSTR reactors treating sewage sludge. Water Res 39:1481–1488
6. Eastman JA, Ferguson JF (1981) Solubilization of particulate organic carbon during the acid phase of anaerobic digestion. J Water Pollut Control Fed 53:352–366
7. Pletschke BI, Rose PD, Whiteley CG (2002) The enzymology of sludge solubilisation utilising sulphate reducing systems. Enzyme Microb Technol 31:329–336
8. Hu D, Lu HP (2004) Placing single-molecule T4 lysozyme enzymes on a bacterial cell surface: toward probing single-molecule enzymatic reaction in living cells. Biophys J 87:656–661
9. Moak M, Molineux IJ (2004) Peptidoglycan hydrolytic activities associated with bacteriophage virions: phage virion murein hydrolases. Mol Microbiol 51:1169–1183
10. Gill AO, Holley RA (2003) Interactive inhibition of meat spoilage and pathogenic bacteria by lysozyme, nisin and EDTA in the presence of nitrite and sodium chloride at 24 °C. Int J Food Microbiol 80:251–259
11. Roman HJ, Burgess JE, Pletschke BI (2006) Enzyme treatment to decrease solids and improve digestion of primary sewage sludge. Afr J Biotechnol 5:963–967
12. Yang Q, Luo K, Li X-M, Wang D-B, Zheng W, Zeng G-M, Liu J-J (2010) Enhanced efficiency of biological excess sludge hydrolysis under anaerobic digestion by additional enzymes. Bioresour Technol 101:2924–2930
13. Ryu DDY, Lee SB (1986) Enzymatic hydrolysis of cellulose: determination of kinetic parameters. Chem Eng Commun 45:119–134
14. Recktenwald M, Wawrzynczyk J, Dey ES, Norrlöw O (2008) Enhanced efficiency of industrial-scale anaerobic digestion by the addition of glycosidic enzymes. J Environ Sci Health A Tox Hazard Subst Environ Eng 43:1536–1540
15. South CR, Hogsett DAL, Lynd LR (1995) Modeling simultaneous saccharification and fermentation of lignocellulose to ethanol in batch and continuous reactors. Enzyme Microb Technol 17:797–803

16. He P-J, Lü F, Shao L-M, Pan X-J, Lee D-J (2007) Kinetics of enzymatic hydrolysis of polysaccharide-rich particulates. J Chin Inst Chem Eng 38:21–27
17. Ferreiro N, Soto M (2003) Anaerobic hydrolysis of primary sludge: Influence of sludge concentration and temperature. Water Sci Technol 47:239–246
18. Mu Y, Wang G, Yu H-Q (2006) Kinetic modeling of batch hydrogen production process by mixed anaerobic cultures. Bioresour Technol 97:1302–1307
19. Li C, Liu G, Jin R, Zhou J, Wang J (2010) Kinetics model for combined (alkaline+ultrasonic) sludge disintegration. Bioresour Technol 101:8555–8557
20. Mu Y, Wang G, Yu H-Q (2006) Kinetic modeling of batch hydrogen production process by mixed anaerobic cultures. Bioresour Technol 97:1302–1307
21. Ferreiro N, Soto M (2003) Anaerobic hydrolysis of primary sludge: influence of sludge concentration and temperature. Water Sci Technol 47:239–246

Chapter 7
Biofuel Generation Process

7.1 Introduction

Increasing global population indirectly increases the world energy demand and way to economic crisis. The continuous burning of petroleum fuel leads to lack of fuel supply and causes several adverse effects on the environment. It is very difficult to balance the increasing global population and their energy demand. Generating bioenergy from renewable feedstock will be the best way to balance the current world energy demand and strengthen the global economic crisis. Biofuels have emerged as a viable source of long-term energy resources and have been classified under renewable energy resources. In transportation, biofuels are blended with existing fuels such as diesel and gas. Blending biofuel with others is an effective way of reducing carbon emission in transportation. Biofuel usage is increasing for the past decade in countries like USA, Brazil, and Europe due to their new energy policies. Around the world, 3% of road transport fuels cover biofuel. It is identified as the most probable, sustainable, and clean energy for the future where it acts as a capable alternate fuel to conventional fuels. Moreover, biofuels are environmentally friendly and during combustion it releases water vapor. Biofuels are having high calorific value and high energy yield. Biofuel generation from biomass is important due to its sustainability and high efficiency. Anaerobic fermentation (dark fermentation and photo fermentation) and electro-microbial hydrogenation (microbial electrolysis cell and microbial fuel cell) are widely used methods to produce biofuels. In this chapter, biofuels such as bioethanol, biohydrogen, and biodiesel generation processes are discussed in detail.

7.2 Bioethanol Generation Process

Bioethanol is an alcohol produced through a fermentative process, chiefly by utilizing carbohydrates present in sugary and starchy plants that include maize, sorghum etc.

© Springer Nature Singapore Pte Ltd. 2022
K. Sudalyandi and R. Jeyakumar, *Biofuel Production Using Anaerobic Digestion*,
Green Energy and Technology, https://doi.org/10.1007/978-981-19-3743-9_7

Cellulose containing biomass has also been established to be the raw material for ethanol generation. Bioethanol could be used as a sole or mixed with traditional liquefied oil to produce gasohol or diesohol. Bioethanol was considered to be an extremely vital fuel in the transport division in a universal outlook, since the generation reached 89.4 billion liters in 2011. A typical reaction of ethanol generation is represented as follows:

$$C_6H_{12}O \rightarrow 2C_2H_5OH + 2CO_2 \qquad (7.1)$$

Generally, yeast and bacteria are extensively employed for bioethanol fermentation. The microbes used in the case of ethanol generation are categorized into three groups viz. yeast (Saccharomyces sp), bacterial strain (Zymomonas sp), and mold (mycelium). Many fermentative processes entail yeast to be the bioagent for transforming the sugary molecules into ethanol in an anaerobic environment.

Starchy materials could not be directly converted to ethanol through the traditional fermentative process. At first, starchy materials should be converted to simpler molecules such as glucose by fermenting it with yeast (saccharomyces cerevisiae, saccharomyces uvarum) to produce bioethanol. Heat tolerant microbes that include Cryptococcus tepidarius and Saccharomyces uvarum have been considered as efficient ethanol-producing bacteria. For generating ethanol through carbon-rich substances such as cane sugar liquor, Zymomonas sp. was considered as an efficient organism that provides better generation higher than 91% [1]. The fermentation time for ethanol generation from waste-activated sludge lasts for 48–72 h. Subsequent to the fermentative process, ethanol is purified through the distillation process. Distillation is to separate two fluids on the basis of their varying boiling points [2].

7.2.1 Bioethanol from Energy Crops and Cellulosic Biomass

The ethanol generated using food crops like maize, paddy etc. is known as cereal ethanol, while ethanol is generated through lignin and cellulose-containing biomass that includes agro waste such as rice and wheat straw and switch grass is called bioethanol. These two ethanols are generated via the biological process. The macromolecular nature of starch limits the production of ethanol by the traditional process. Firstly, the polymeric chain has to be broken down to simpler molecules of glucose prior to the fermentation process. During this process, starchy raw materials are pulverized and blended with water to form a smash that characteristically holds 16–21% starchy materials. Smash is boiled afterwards on or over its steaming point and subsequently subjected to treatment by two enzymes namely, amylase and pullulanase. At first, the enzyme, amylase converts the starchy materials to smaller molecules. Through hydrolysis, amylase releases the oligosaccharides "maltodextrin" from starch [3]. The released maltose and dextrin are further degraded through enzymes, pullulanase, and glucoamylase through saccharification. Saccharification transforms the entire dextrins to simple sugars such as glucose and maltose. Smash

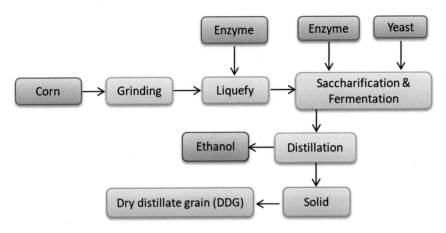

Fig. 7.1   Dry mill process

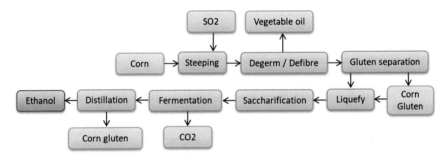

Fig. 7.2   Wet mill process

is again chilled to 29 °C and was inoculated with yeast to facilitate fermentation for the production of ethanol. Ethanol generation amenities from maize could be categorized into wet and dry mill processes. Dry mills are typically very small in magnitude (capability) and are designed mainly to generate ethanol alone. The starchy cereals are processed for ethanol generation through wet milling and dry grinding, and their process flow diagram is shown in Figs. 7.1 and 7.2. Wet grinding ethanol fermentation generates various expensive co-products that include organic acids and solvent.

Dry mill process is particularly considered in the case of generation of ethanol and animal feed.

7.2.2   Bioethanol from Lignocellulosic Biomass

Lignin, cellulose, and starch containing biomass are subjected to hydrolysis (microbial or enzymatic-based biochemical transformations) to produce sugars. These

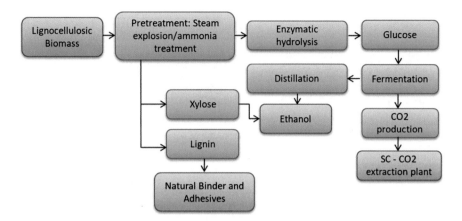

Fig. 7.3 Generation of ethanol from lignocellulosic biomass

sugars are then transformed to ethanol and various biofuels. Among the biological conversion process, yeast-based fermentation of sugar and starch to ethanol is efficient. Figure 7.3 shows the process flow diagram of ethanol generation from lignocellulosic biomass. The transformation of biomass raw materials to ethanol involves various processes such as the disintegration of raw material, hydrolytic processes, fermentative processes, and ethanol upturn (Fig. 7.3). The pretreatment process such as steam explosion and ammonia treatment disintegrates xylose and lignin of cellulose. The steam explosion technique is considered to be the effective pretreatment process for transforming lignin and cellulose containing materials.

During this technique, the substances to be disintegrated are kept in a reactor and subjected to exposure of saturated steam for a period of 20 s to 20 min at a temperature of 245–283 K and elevated pressure of 13–17 bar. The pressure that exist in the reactor is subsequently reduced through releasing the steam. Then the substance is subjected to normal environmental pressure to form an explosion that breaks up lignin and cellulose containing materials [2]. Various reactions are involved in steam explosion. Firstly, steam explosion leads to the decomposition of hemicellulose and lignin. Secondly, these substances got transformed into smaller particle size substances and can be removed out of the system. As a result, liquefied portion of hemicellulose could easily be extracted through water removal.

The xylose could be converted to ethanol and lignin could be additionally treated to generate various biofuels [4]. The cellulosic leftovers such as solid substances after disintegration again disintegrated to yield glucose through enzymatic treatment. The glucose is again converted to ethanol by fermentative process, and the hemicellulose portion is transformed to xylose. The transformation of xylose to ethanol is a complex technique; hence, pretreatment is essential.

Table 7.1 List of typical composition of syngas

Typical composition of syngas		
Component	Formula	Value (%)
Carbon monoxide	CO	49.98
Hydrogen	H_2	32.42
Carbon dioxide	CO_2	7.2
Nitrogen	N_2	4.52
Methane	CH_4	3.9
Other gases	–	1.98

7.2.3 Bioethanol from Syngas

Syngas, a blend of carbonmonoxide, hydrogen, and various hydrocarbons that can be generated through incomplete burning of feedstock, i.e. during burning, the quantity of O_2 was inadequate to transform the biomass entirely to carbon dioxide and water. Table 7.1 lists the typical composition of syngas. Major amount of energy present in syngas can be removed and used for various purposes. Syngas can be burned in internal combustion engine and heat energy chambers [3]. Syngas is also used for the production of ethanol through fermentation. During fermentation, the anaerobic microbes consume these gases and produce ethanol.

7.2.4 Syngas Fermentation

Syngas fermentation is a complex transformation process for the generation of ethanol and various biofuels. In contrast to anaerobic fermentation, syngas fermentation is considered to be an oblique process since the raw materials have not been openly loaded into the reactor allowing the process to occur. Firstly, raw materials have to be gasified to syngas. Secondly, it is purified and chilled prior to be loaded to the reactor for product formation. Non food related raw materials that include agro wastes, municipal solid wastes, energetic plants, and petroleum hydrocarbons could be gasified to generate syngas. Syngas composition mainly relies upon the nature of raw materials and the type of gasification process. In syngas fermentation, tars could taint fermentation vessels and various pollutants such as nitric oxide, hydrogen sulphide, and hydrogen cyanide could hamper microbial enzymatic action. The catalyst mediated the transformation of syngas to ethanol through Fisher–Tropsch (FT) process that was developed in 1923 by German scientists F. Fisher and H. Tropsch. During 1987, a bacilli form gram-positive anaerobic bacterium Clostridium ljungdahlii was identified that was recognized to possess the ability of fermenting CO and H_2 to ethanol and acetic acid [5].

Figure 7.4 shows the biochemical pathway of converting syngas to ethanol which is also called as acetyl CoA pathway or Wood Ljungdahl pathway (named after the scientists, Harland Goff Wood and Lars Gerhard Ljungdahl who invented this pathway). In the biochemical pathway of syngas to ethanol, acetogenic bacteria metabolize single carbon compounds. It involves three steps. (i) it forms acetyl moiety of acetyl CoA from carbondioxide, (ii) stores energy, (iii) consumes carbondioxide, and converts them to cell carbon. This pathway involves both oxidation and reduction reactions. Carbon dioxide gets converted into acetate through a reduction reaction. The formed acetate is again converted back into carbon dioxide through an oxidation reaction. The formation of acetyl CoA in this pathway involves two steps namely, methyl step and carbonyl step. The formed acetyl CoA then gets converted into acetate, ethanol, and cell mass.

The conversion of acetyl CoA to acetate is called acetogenesis and the conversion of acetate to ethanol is called solventogenesis. The oxidation of hydrogen by hydrogenase and oxidation of carbon monoxide by carbon monoxide dehydrogenase (CODH) supplies the required electrons for reduction reactions as shown in Fig. 7.4. The electrons formed from oxidation of hydrogen and carbon monoxide are carried by the electron carriers such as NADH/NAD+, NADPH/NADP+, or ferredoxin. During this electron transfer, adenosine tri phosphate (ATP) supplies chemical

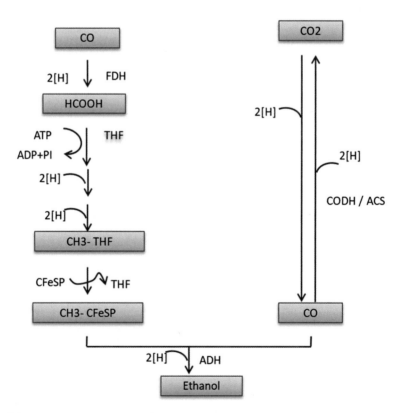

Fig. 7.4 The biochemical pathway of converting syngas to ethanol

energy for cell metabolism [6]. The ATP gets converted into ADP through hydrolysis of phosphate bonds and the release of energy takes place.

The general biochemical reactions for the transformation of syngas to ethanol and acetic acid are summarized as follows:

$$6CO_2 + 3H_2O \rightarrow C_2H_5OH + 4CO_2 \ [21] \tag{7.2}$$

$$2CO_2 + 6H_2 \rightarrow C_2H_5OH + 4CO_2 \tag{7.3}$$

$$4CO_2 + 2H_2O \rightarrow CH_3COOH + 2CO_2 \tag{7.4}$$

$$2CO_2 + 4H_2 \rightarrow CH_3COOH + 2H_2O \tag{7.5}$$

7.3 Biohydrogen

Hydrogen is an energy-rich and non-polluting biofuel when compared with other fuels. It could be generated from renewable sources through various techniques. These features formulate H_2 as an attractive fuel meant for upcoming energy process on the basis of sustainability [7]. Hydrogen possesses tremendous probable characteristics to offer bioenergy by reducing global warming and environmental production. Hydrogen in addition is considered to be excellent for in-situ ignition machines and upholding durable instrument effectiveness. The biohydrogen generation from microalga and bacteria has begun 49 years ago during the meeting held on the topic "Biological energy conservation". In 1973, the evolution of hydrogen from water mediated by the enzyme hydrogenase (synthesized by the bacterial sps *Clostridium kluyveri*) was investigated by Bennman and Coworkers. He and his co-workers haved mixed chloroplast extracted from spinach with the enzyme hydrogenase synthesized by *Clostridium kluyveri* and ferredoxin and discovered the evolution of hydrogen. These results revealed that the evolution of hydrogen from water by photosynthetic means in the presence of sunlight was considered as the solar energy conversion technique. Researchers have deployed the production of hydrogen from spinach plants photosynthetically in the mid-2000. However, the production of hydrogen in a biological way was endured as lab scale trials. Production of hydrogen from biomass or waste materials via dark fermentation affords an alternate means to biological hydrogen production. Biohydrogen production by employing anaerobic microbial population has been investigated since 1980s. Production of biohydrogen from sunlight through direct mode provides greater theoretical efficiency than photosynthetic efficiency of crops. On the other hand, the economic viability of the method demands more advancement as the cost associated with photobioreactors was a major limitation. Though biohydrogen production is the sequential process mediated by microbes, the overall reaction step can be simplified into the following single reaction. This reaction can be mediated by hydrogenase or nitrogenase enzyme

$$2H^+ + 2e \rightarrow H_2. \tag{7.6}$$

Biohydrogen generation is considered to be a green process, since it reduces the usage of fossil fuels. The carbon that has been discharged via hydrogen is got by means of carbon dioxide fixation and through microbial fermentation. Therefore, biological hydrogen could be regarded as carbon less. Biohydrogen is considered to be a feasible resource that provides the present energy requirement and eliminates ecological problems.

7.3.1 Process and Metabolic Routes of Biohydrogen Production Methods

Biohydrogen production via biological way is existing with reference to acidogenic or anaerobic fermentative process or dark fermentative process, biophotolysis, enzymatic, and electrogenic routes. The biological hydrogen production process can be categorized further into light-dependent (photosynthetic) and light-independent (dark) fermentation processes based on the dependence of light. By another means, the photobiological hydrogen production processes can be further categorized into either fermentation or photosynthetic process based on the source of carbon and microbes or enzyme (biocatalyst) employed. The photosynthetic mode of the process can happen via direct or indirect biophotolysis of water mediated by green alga and cyanobacteria or through photofermentation mediated by photosynthetic bacteria (PSB). The main benefit of biophotolysis is that it effectively converts solar energy into hydrogen. The theoretical energy efficiency of direct biophotolysis process is approximately 40%. During direct and indirect biophotolysis, microalgae and cyanobacteria utilizes inorganic carbondioxide to generate hydrogen under sunlight and water. Photosynthetic bacteria mediate photofermentation by utilizing various substrates (both organic and inorganic) under sunlight. Dark fermentation process refers to anaerobic metabolism in which anaerobic microbes (majorly acidogenic population) produce hydrogen together with volatile fatty acids and carbondioxide metabolically via acetogenesis. During electrogenesis, an external potential is employed to the microbial cells to enhance biohydrogen generation. A suitable biocatalyst mediates the process electrochemically during which the biocatalyst combines two electrons from oxygen generating phase with two H^+ to synthesize hydrogen.

Several other processes such as biogasification and electrolytic processes could produce hydrogen, below 20 USD/GJ, that is fairly realistic [8]. Hydrogen possesses the greater energy content/unit weight (143 kilojoule per gram) of the recognized biofuels, though it needs unique management as slight energy is required in case of combustion. Hydrogen is considered to be a potent energy resource as it generates only water while combustion. Various digesters have been presently employed to generate biohydrogen through fusion, electrochemical, and catabolic processes. The major drawback in industrializing biological hydrogen is the marginal generation and reduced rate of generation [9]. Utilizing inexpensive feedstocks, proficient generation processes, and pilot scale examination of photofermentation units must form biological hydrogen a potent feasible resource of bioenergy in future. In this section, the following hydrogen generation processes are detailed.

- Dark fermentation Process
- Photo fermentation Process
- Biophotolysis Process.

7.3.2 Dark Fermentation Process

Dark fermentation happens under anaerobic environment, where hydrogen is produced via the decomposition of organic materials by anaerobic microbes [10]. The organic compounds serve as an energy source for the generation of hydrogen. The hydrogen generation from the various substrate was provided in Table 7.2. During the dark fermentation process, mixture of gases such as hydrogen, carbondioxide, methane, carbonmonoxide, and hydrogen sulphide has been generated, and their composition depends upon the type of hydrogen generation process and characteristics of raw materials [11]. As given in the below equation, the absolute oxidation of glucose to hydrogen and carbondioxide generates the utmost twelve moles of hydrogen for one mole of glucose. The energy required for this catabolic process is nil.

$$C_6H_{12}O_6 + 6\,H_2O \rightarrow 12H_2 + 6CO_2 \quad \Delta G0 = +3.2\,kJ \tag{7.7}$$

$C_6H_{12}O_6$, isomers of 6 carbon monomeric compound, and polymers such as starch and cellulose, generates a varying quantity of hydrogen per mole of glucose, relying upon fermentative reactions and final products [12]. The active hydrogen generation from glucose is estimated through butyric acid to acetic acid ratio. When acetic acid is the terminal product, a higher theoretical product of 4 mol H_2/mole $C_6H_{12}O_6$ is obtained [13].

$$C_6H_{12}O_6 + 2H_2O \rightarrow 4H_2 + 2CH_3COOH\ (acetate) + 2CO_2\ \Delta G0 = -206\,kJ \tag{7.8}$$

At the same time while butyric acid is a terminal product, a higher theoretical yield of 2 mol H_2/mole $C_6H_{12}O_6$ is generated.

$$C_6H_{12}O_6 + 2H_2O \rightarrow 2H_2 + CH_3CH_2CH_2COOH\ (butyrate)$$

Table 7.2 Hydrogen generation from various substrates

Substrate	Microorganism/Reactor type	Organic products in fermentation broth	Conditions: pH/Temp	Hydrogen productivity/yield	References
Organic municipal solid waste 110 g TVS/dm^3/d	CSTR/Mixed cultures semi-continuous	Butyric acid, acetic acid	pH = 5.0 T = 50 °C	5.7 dm^3 H$_2$/dm^3/d	[28]
Kitchen garbage	Sludge from anaerobic digesters/CSTR	Acetic acid, ethanol, lactic acid	pH = 5.0 T = 55 °C	1.7 dm^3 H$_2$/dm^3/d 66 cm^3 H$_2$/g VS 12.5 mmol H$_2$/dm^{3}h	[29]
Potato steam peels 10 g glucose/dm^3	Mixed culture/Batch	Acetic acid, lactic acid	pH = 6.9 T = 75 °C	3, 8 mol H$_2$/mol glucose	[30]
Simulated food waste: Fish 5%; meat 10%; bread 10%; apple 10%; kiwi 6%; banana 9%; pear 10%; onion 5%; lettuce 5%; carrot 5%; cabbage 10%; potato 15%	CSTR continuous/mixed culture from digested sludge	Acetic acid, butyric acid, caproic acid, valeric acid	pH = 5.5 T = 34 °C	20.5 dm^3 H$_2$/kgVS	[31]
Liquid swine manure 13.94 g COD/dm^3	ASBR Batch of mixed cultures from anaerobic digesters	Acetic acid, butyric acid, valeric acid, ethanol, propionic acid	pH = 5.0 T = 37 °C	0.1 dm^3 H$_2$/dm^3/h	[32]
Cattle wastewater 1.3 g COD/dm^3	Sewage sludge/Batch	Butyric acid, acetic acid, ethanol, propionic acid	pH = 5.5 T = 45 °C	0.34 dm^3/dm^{3}h	[13]
Cornstalk	Anaerobic sludge that has been heat pretreated	Acetic acid, butyric acid	pH = 4.8 T = 50 °C	126.22 mL H$_2$ g^{-1} DB	[7]
Cornstalk	Anaerobic sludge that has been heat pretreated	Acetic acid, butyric acid	pH = 4.8 T = 50 °C	141.29 mL H$_2$ g^{-1} DB	[33]

(continued)

Table 7.2 (continued)

Substrate	Microorganism/Reactor type	Organic products in fermentation broth	Conditions: pH/Temp	Hydrogen productivity/yield	References
Cornstalk	Aerated microbial *Consortium*	Acetic acid, butyric acid	pH = 4.8 T = 50 °C	176 mL H_2 g^{-1} DB	[8]
Beer lees	Anaerobic mixed *Consortia*	Acetic acid, butyric acid		53.03 mL H_2 g^{-1} DB	[9]
Lawn grass	Enriched mixed culture dominated by *C. pasteurianum*	Acetic acid, butyric acid		72.21 mL H_2 g^{-1} DB	[10]
Soybean Straw	Enriched mixed culture dominated by *C. butiricum*	Acetic acid, butyric acid		60.2 mL H_2 g^{-1} DB	[11]
Cornstalk	*T.thermosaccharolyticum W16*	Acetic acid, butyric acid	pH = 5.5 T = 29 °C,	89.3 mL H_2 g^{-1} DB	[12]
Poplar Leaves	Enriched mixed culture dominated by *C. pasteurianum*	Acetic acid, butyric acid	pH = 4 T = 50 °C	44.92 mL H_2 g^{-1} DB	[34]
Cornstalk	Enriched consortia from cow dung compost	Acetic acid, butyric acid		149.69 mL H_2 g^{-1}	[14]
Sugarcane Bagasse	*C. butyricum*	Acetic acid, butyric acid	pH = T = 121 °C	44 mL H_2 mmol^{-1}	[15]
Reed canary grass (RCG)	Enriched microbial culture	Acetic acid, butyric acid	pH = T = 121 °C	31.6 mL H_2 g^{-1} DB	[16]
Miscanthu	*T. elfii DSM 9442*	Acetic acid, butyric acid	pH = 4.8 T = 45 °C	2280 mL H_2	[35]
Rice straw	*T. neapolitana*	Acetic acid, butyric acid	pH = T = 121 °C	77.1 mL H_2 g^{-1} DB	[36]
Sweet sorghum bagasse	*C. saccharolyticus*	Acetic acid, butyric acid	pH = 5 T = 50 °C	73.6 mL H_2 mmol^{-1} C_6sugarse	[37]

$$+\ 2CO_2 (\Delta G0 = -254\,kJ) \tag{7.9}$$

Therefore, a higher theoretical product of hydrogen is associated to acetic acid as a fermentative terminal product. In point of fact, greater hydrogen generation is associated to the mixture of acetic acid and butyric acid fermentative production, at the same time less hydrogen generation is related to propionic acid and minimized terminal products, for instance ethanol and lactic acid. The liquefied metabolic products resulting through the dark fermentative process, comprising of volatile fatty acids and alcohols, were additionally utilized for hydrogen generation in consequent photofermentation as described in the reaction as follows:

$$2CH_3COOH + 4H_2O \rightarrow 8H_2 + 4CO_2 \tag{7.10}$$

7.3.3 Metabolic Routes of Dark Fermentative Biohydrogen Production

Figure 7.5 represents the metabolic route of dark fermentative hydrogen production. In the dark fermentative metabolic route for biohydrogen production, three biochemical reactions are involved. The first metabolic route is studied in the bacterium, *E. coli*, and *Enterobacteriaceae*. Two major enzymes are involved in this pathway namely, Pyruvate formate lyase (PFL) and formate hydrogen lyase (FHL). In this metabolic route, the pyruvate, which is formed through glycolic pathway, is cleaved into acetyl-CoA and formate under anaerobic conditions. This reaction is mediated by the enzyme PFL. The formed formate is then splitted into hydrogen and carbondioxide by FHL as represented in the following Eqs. (7.11) and (7.12):

$$\text{Pyruvate} + \text{CoA} \xrightarrow[\text{PFL}]{} \text{Acetyl CoA} + \text{Formate} \tag{7.11}$$

$$\text{H}_2 + \text{CO}_2 \xrightarrow[\text{FHL}]{} \text{Formate} \tag{7.12}$$

The two enzymes that are involved in second metabolic route of hydrogen production are pyruvate: ferredoxin oxidoreductase (PFOR) and Fd-dependent hydrogenase (HydA). The pyruvate from glycolytic pathway undergoes oxidative decarboxylation and converted into acetyl CoA and carbondioxide under anaerobic conditions. This biochemical reaction is mediated by the enzyme PFOR.

In this metabolic route, the electrons are first transported to Fd_{ox} (oxidized ferredoxin). Fd_{ox} is an electron acceptor that have greater negative potential (E_o—420 mV). The electrons in Fd_{rd} (reduced ferredoxin) are converted to protons to generate hydrogen. This biochemical reaction is mediated by the enzyme Hyd A as described in following Eqs. (7.13) and (7.14). This type of metabolic route is distinctive for bacterial species, *Clostridium*.

$$\text{Pyruvate} + \text{CoA} + \text{Fd}_{ox}\text{Acetyl} \xrightarrow[\text{PFOR}]{} \text{CoA} + \text{CO}_2 + 2\,\text{Fd}_{rd} \tag{7.13}$$

$$2\,\text{Fd}_{rd} + 2\text{H}^+ \xrightarrow[\text{Hyd A}]{} 2\,\text{Fd}_{ox} + \text{H}_2 \tag{7.14}$$

In third type metabolic route, NAD(P)H is utilized to generate hydrogen. The two major enzymes that are involved in this biochemical reaction are NAD(P)H ferredoxin oxidoreductase (NFOR) and HydA. Then, the NAD(P)H which is generated in carbon metabolism reduces Fd_{ox}. The electrons present in the Fd_{rd} are converted to protons. These protons are converted to hydrogen and this reaction is mediated by the enzyme Hyd A and the reactions are represented in following Eqs. (7.15), (7.16) and (7.17). This biochemical reaction is typical in many thermophiles and few *Clostridium sps.*

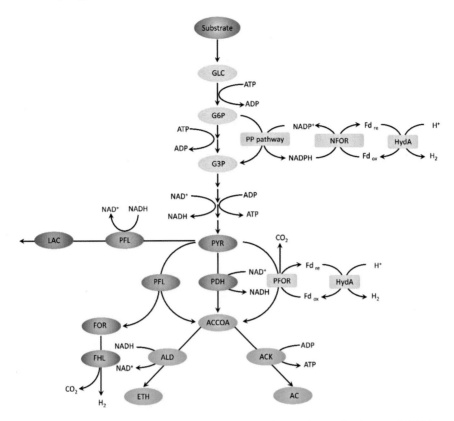

Fig. 7.5 Metabolic route of dark fermentative pathway (Abbreviations: AC—Acetate; ACCOA—Acetyl CoA; ACK—Acetate kinase; ALD—Alcohol dehydrogenase; ETH—Ethanol; Fdox—Oxidised Ferredoxin; Fdox—Reduced ferredoxin; FHL—Formate hydrogen lyase; FOR—Formate; GLC—Glucose; G3P—Glyceraldehyde 3 phosphate; G6P—Glucose-6—phosphate; Hyd A—Ferredoxin dependent hydrogenase; LAC—Lactate; LDH—Lactate dehydrogenase; PP Pathway—Pentose phosphate pathway; NFOR, NADP (H)—Ferredoxin Oxidoreductase; PDH—Pyruvate dehydrogenase; PFL—Pyruvate formate lyase; PFOR—Pyruvate ferredoxin oxidoreductase; PYR—Pyruvate)

$$\text{Glucose} + 2\text{NAD}^+ \xrightarrow[\text{Glycolytic pathway}]{} 2 \text{ Pyruvate} + 2 \text{ NADH} \qquad (7.15)$$

$$2\text{NADH} + 4 \text{ Fd}_{\text{ox}} 2 \text{ NAD}^+ \xrightarrow[\text{NFOR}]{} +4 \text{ Fd}_{\text{rd}} \qquad (7.16)$$

$$4 \text{ Fd}_{\text{rd}} + 4 \text{ H}^+ 4 \text{ Fd}_{\text{ox}} \xrightarrow[\text{Hyd A}]{} +2\text{H}_2 \qquad (7.17)$$

7.3.4 *Photofermentation*

Photofermentation is the process of transformation of organic materials to biohydrogen mediated through different classes of photosynthetic bacteria which involves many biological steps as like that of anaerobic digestion [14, 15]. Photo fermentative process varies from dark fermentative method as it takes place in the presence of solar energy. Photo fermentative process is performed by oxygenic photosynthetic bacteria which utilizes light and biomass waste to generate hydrogen [16]. Among the bacterial population, purple non sulfur and green sulfur bacteria are proficient in generating H_2 through utilizing light power and minimal by products. Their photo systems vary when compared to oxygenic photosynthesis because of need for food and their incapability to oxygenate water.

(a) **Metabolic route of photofermentative pathway**

Figure 7.6 represents photofermentative metabolic route in purple sulfur bacteria. The electron donors are sulfide or organic substrate in photofermentative metabolic pathway. In this pathway, reduction of NAD^+ to NADH takes place via reversed electron flow. Then the electrons are transported via ferredoxin to nitrogenase enzyme in which consumption of ATP and production of hydrogen takes place. 4 ATP molecules are required to produce 1 mol of hydrogen. In case of purple non-sulfur bacteria (Fig. 7.7), the produced hydrogen supply electrons for photosynthetic reaction.

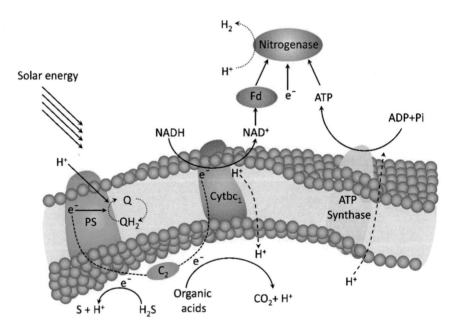

Fig. 7.6 Photofermentation in purple sulfur bacteria

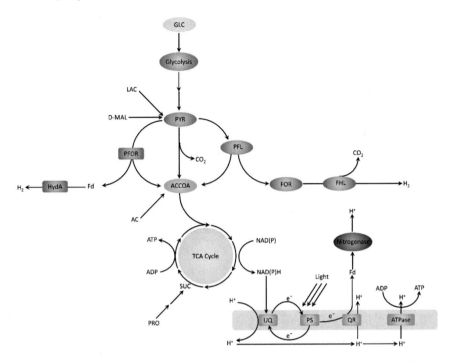

Fig. 7.7 Photofermentation in non purple sulfur bacteria (Abbreviations: AC—Acetate; ACCOA—Acetyl CoA; ATPase—ATP synthase; D MAL—D malate; Fd—Ferredoxin; FHL—Formate hydrogen lyase; FOR—Formate; GLC—Glucose; Hyd C—Ferredoxin dependent hydrogenase; LAC—Lactate; N_2ase—Nitrogenase; OR—Oxidoreductase; PFL—Pyruvate formate lyase; PFOR—Pyruvate ferredoxin oxidoreductase; PRO—Propionate; PS—Photosystem; SUC—Succinate; UQ—Ubiquinone)

During this pathway, the carbondioxide is fixed through calvin cycle and O_2 serve as terminal electron acceptor. In anaerobic condition, the scavenged electrons from organic acids are transferred to Fd_{ox}. The electrons are transferred via a sequence of membrane bound electron transport carriers. Then the electrons from Fd_{ox} reduces molecular nitrogen to ammonia. This reaction is mediated by the enzyme nitrogenase as represented in Eq. (7.18). In case of nitrogen free system, nitrogenase enzyme mediates the reduction of protons to synthesize hydrogen as represented in Eq. (7.19).

$$N_2 + 8\,H^+ + 8\,e^- + 16\,ATP \xrightarrow[\text{Nitrogenase}]{} 2NH_3 + H_2 + 16\,ADP + 16\,pi \quad (7.18)$$

$$8\,H^+ + 8\,e^- + 16\,ATP \xrightarrow[\text{Nitrogenase}]{} 4H_2 + 16\,ADP + 16\,pi \quad (7.19)$$

An electrochemical ingredient is generated across the membrane bound electron carrier during photophosphorylation and this gradient is utilized for the synthesis of ATP. The proton or nitrogen reduction needs enormous amount of energy (ATP) to overwhelm the elevated activation energy. In addition two ATP molecules are needed to move electrons from Fe protein to nitrogenase. Oxygen is toxic to nitrogenase enzyme and as a result, the production of hydrogen may get suppressed. Therefore, hydrogen production by purple nonsulfur bacteria needs oxygen free system.

(b) **Biophotolysis**

Production of biological hydrogen through blue green and green algal biomass by photolysing water with the help of sun light and enzymes such as hydrogenase and nitrogenase is called as biophotolysis [17, 18]. Microalgal biomass utilizes solar energy for transferring electrons to nicotinamide adenine dinucleotide phosphate hydrogenase and ferredoxin, that consecutively produces hydrogen. To recommend an inexpensive reasonable resource of hydrogen, the algal biomasses should attain as a minimum of 9% solar transformation effectiveness; the green alga *Chlamydomonas reinhardtii* attained a higher theoretical light transformation effectiveness of 21% in optimal lab condition [19–23]. There are two types of biophotolysis.

- Direct photolysis
- Indirect photolysis.

Direct photolysis

During direct photolysis, H_2O molecules splitted to form hydrogen and oxygen in the presence of solar energy. The reaction is represented as below:

$$2H_2O + \text{light energy} \rightarrow 2H_2 + O_2 \qquad (7.20)$$

Photosystem I and II are responsible for photosynthesis process in microalgal biomass. Figure 7.8 shows the process scheme of direct photolysis. In direct photolysis process, when photosystem I and II were hit by solar energy it emit electrons [24, 25]. The emitted electrons are transported to ferredoxin, utilizing solar energy taken up by photosystem. Hydrogenase subsequently takes up electron of ferredoxin to produce hydrogen.

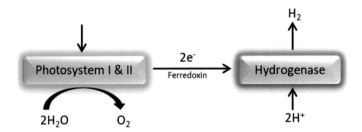

Fig. 7.8 Direct photolysis process

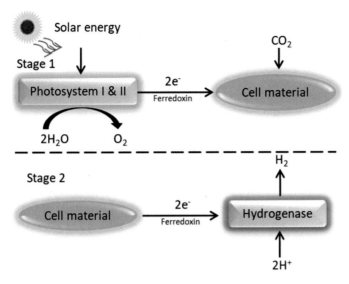

Fig. 7.9 Indirect photolysis process

Indirect photolysis

It is a two step process (Fig. 7.9). In the first step, excited electrons of photosystem I and II are transported to cell material through ferredoxin [26]. In the second step, cell material undergoes fermentation and produces two electrons, which are subsequently converted into hydrogen through hydrogenase [27]. The following reaction occurs during this process.

$$12H_2O + 6CO_2 \rightarrow C_6H_{12}O_6 + O_2 \tag{7.21}$$

$$C_6H_{12}O_6 + 12H_2O \rightarrow 12H_2 + 6CO_2 \tag{7.22}$$

7.3.5 Diversity of Microbes in Biohydrogen Producing System

The microbes exist in the hydrogen producing system are classified into two major types namely, hydrogen producing and non-hydrogen producing microbes. The hydrogen producing microbes are further classified as photosynthetic, photo fermentative and dark fermentative microbes based on their metabolism in synthesizing biohydrogen. The photosynthetic microbes are cyanobacteria and green algae. These microbes split water into H_2 and O_2 in the presence of sunlight. The photofermentative microbes are purple sulfur bacteria (*Chromatium*), purple non sulfur bacteria

(*Rhodobacter*), green sulfur bacteria (*Chlorobium*) and gliding bacteria (*Chloroflexus*). These microbes utilize organics and short chain volatile fatty acids using sunlight as energy source to produce hydrogen. The dark fermentative microbes are extensively present microbes. These microbes include *Clostridium* and *Enterobacter sps*. These microbes also include bacteria that present in extreme habitats such as thermophilic hot spring (e.g., *Thermoanaerobacterium sps*) and the psychrophilic polar areas (e.g., *Polaromonas sps*). These microbes utilize organics as substrate and produce hydrogen through sequential biochemical reaction steps.

Dark fermentation can be performed in the absence of light. In addition to hydrogen producing microbes, the other microbes that are present in the system are the mixed microbial population (inoculum). Among these microbes, some are essential and some are unwanted non-hydrogen producing microbes. The non-hydrogen producing microbes are hydrogen consuming bacterial populations such as methanogen and homoacetogenic bacteria and the substrate competitive microbes. The existence of these unwanted non-hydrogen producing microbes lowers the hydrogen production and yield in the system. These unwanted microbes can be removed via inoculum pretreatment and control of operation.

In addition, the existence of non-hydrogen producing microbes may also have some advantages as it possess multiple metabolic routes for the biodegradation of complex waste and enhances the biohydrogen production. For instance, some bacteria such as *Streptococcus sps*. can enhance the biohydrogen generation through formation of granules and biomass retention. Besides, the existence of some aerobic or facultative anaerobic microbes can aid in maintaining anaerobic environment in the system.

Inoculum for dark fermentation

The two major types of inoculum used for dark fermentative hydrogen generation are mixed cultures (e.g., anaerobically digested sludge, compost, leachate, soil etc.) and pure cultures (e.g., *Clostridium sps, Enterobacter sps*. etc.). The mixed cultures are the most extensively employed microbes due to wider option of substrate, cost effective system function and ease regulation of the system. Using co culture of various microbes in the system are more efficient in hydrogen production in case of complex feedstocks. Using pure cultures in the system demands more operational and maintenance cost. On the other hand employing pure cultures can leads to thorough knowledge about the metabolic routes in hydrogen production. Isolation and identification of efficient hydrogen producing strains with special gene of interest (will be discussed in following section) is the effective way to enhance the hydrogen production and yield.

Mixed Cultures

Mixed cultures are the diverse of microbes that are extensively present in natural habitats such as anaerobically digested sludge, compost, sediments, soil leachate, organic waste etc. These are the possible sources for enrichment of hydrogen producing microbes. Among these sources, anaerobically digested sludge is the widely employed resource of hydrogen producing microbes. In case of employing organic wastes such as waste activated biosolids, municipal solid waste etc., the native microbes could be employed as hydrogen producing microbes and therefore extra inoculum was not needed. A diverse group of microbes present in inoculum. The dominant microbes in anaerobically digested sludge are mostly *Clostridium sps* which includes *C. butyricum, C. pasteurianum* and *C. beijerinckii*. The most common dominant microbes in compost are *Enterocbacter sps, Bacillus* and *Enterococcus sps* coexist with *Clostridium sps*.

Pure Cultures

Diverse variety of microbes can be employed to produce hydrogen. The most commonly employed pure cultures are *Clostridium sps* which include *C. butyricum, C. beijerinckii, C. pasteurianum, C. tyrobutyricum* etc. and *Enterobacter sps* which include *E. aerogenes, E. asburiae, E. cloacae*, etc. Other examples of pure cultures for potential biohydrogen production are *Ethanoligenens, Bacillus, Klebsiella, Thermoanaerobium, Rahnella* etc. The pure culture hydrogen producing microbes are classified as psychrophiles (0–20 °C), mesophiles (30–40 °C) and thermophiles (50–100 °C) based on the growth temperature. The examples of psychrophiles are *Rahnella sps, Polaromonas sps* etc. The examples of mesophiles are *Clostridium sps, Enterobacter sps* etc. The examples of thermophiles are *Klebsiella sps, Thermoanaerobium sps*. etc. Mesophiles are mostly preferred for biohydrogen production as they bring cost effective biohydrogen production. The pure cultures for biohydrogen production are further classified as anaerobes, and facultative anaerobes based on their oxygen tolerance level. The example for anaerobes and facultative anaerobes is *Clostridium sps* and *Enterobacter sps*. The biohydrogen production potential of anaerobes are greater. Facultative anaerobes aid in maintaining anaerobic environment for anaerobes in the hydrogen production system.

7.3.6   *Enzymes Involved in Biohydrogen Production*

The enzymes involved biohydrogen production are hydrogenase and nitrogenase. The main enzyme that catalyzes the synthesis of hydrogen from protons or oxidation of hydrogen to protons are called hydrogenase. The biochemical reaction during hydrogen production catalysed by hydrogenase is represented as follows:

$$2H^+ + 2e \rightarrow H_2.$$

This biochemical reaction is reversible. The direction of this reaction is based on the oxido reduction potential of the substrates which are able to react with the enzyme hydrogenase. The hydrogenase enzyme plays a key role in regulating the metabolic pathway of hydrogen producers and in enhancing hydrogen. Nitrogenase is another enzyme that usually mediates the conversion of nitrogen to ammonia via reduction reaction. Under photo heterotrophic environment, nitrogenase enzymes are capable of reducing protons into hydrogen.

Hydrogenase enzymes and its classification

The hydrogenase enzymes are classified as oxygen sensitive, oxygen resistant and oxygen tolerant enzymes. The oxygen sensitive enzymes undergo irreversible inactivation when exposed to oxygen. The oxygen resistant enzymes undergo suppression and can be recovered under anaerobic environment. The oxygen tolerant enzymes are active in aerobic condition and mediates the oxidation of hydrogen. Certain hydrogenase enzyme mediates the reversible hydrogen oxidation and synthesis. Some enzymes can mediate only either hydrogen synthesis or oxidation. Some hydrogen producing microbes have extra hydrogenase enzymes and their function differs. The hydrogenase enzymes are classified into three forms such as [Fe]$^-$, [FeFe]$^-$, and [NiFe]$^-$hydrogenases based on their metal content in active site.

[Fe]$^-$Hydrogenases

[Fe]$^-$Hydrogenase is also known as hydrogen forming methylene tetrahydromethanopterin dehydrogenase (HMD) and iron sulfur cluster free enzyme as these does not contain clusters of iorn sulfur in their structure. These enzymes consist of three clusters. The active site of this enzyme has iron center. The iron center coordinates with atom of cysteine sulphur, two cis-co ligands, a bidentate pyridine via nitrogen and carbon atoms. This enzyme converts methenyltetrahydromethanopterin along with hydrogen to methylene tetrahydromethanopterin via reversible reduction reaction. This reversible reduction reaction is the intermediate step during reduction of carbon dioxide into methane mediated by certain methanogens. During this reaction, a hydride from hydrogen is moved to the pro-R location of carbon methylene group of the formed product.

[NiFe]$^-$Hydrogenases

These enzymes are the important class of hydrogenase. These enzyme converts molecular hydrogen into two protons and electrons through heterolytic cleavage reaction. This enzyme is also capable to mediate production of hydrogen from two protons and electrons through reduction reaction. In these enzymes, the transfer of electrons and protons takes place between the catalytic center and the molecule surface. In Knallagas bacterium, *Ralstonia eutropha*, these enzymes are connected to the respiratory chain. Therefore, this bacterium possesses greater resistance to O$_2$ and CO.

[FeFe]$^-$Hydrogenases

These enzymes mediate the interconversion of hydrogen with electrons and protons. The active site of this enzyme is bi octahedral in structure and it consists of two iron cores. The active site is also known as H cluster. The active site is connected to FeS cluster. The FeS cluster is located in the N terminal end of the active site and this cluster is accountable for the reversible transfer of electron from the active site. These enzymes can be classified into two sub classes namely, (a) Cytoplasmic, soluble monomeric [FeFe]-hydrogenases (b) Periplasmic, heterodimeric [FeFe]-hydrogenases.

(a) Cytoplasmic, soluble monomeric [FeFe]-hydrogenases
     These enzymes are present in bacterium such as *Clostridium pasteurianum* and *Megasphaera elsdenii*. These enzymes mediate hydrogen production and utilization reaction. These enzymes are sensitive to oxygen and they are found in strict anaerobic bacteria.
(b) Periplasmic, heterodimeric [FeFe]-hydrogenases
     These enzymes are present in the bacterium such as *Desulfovibrio spp.* These enzymes mainly mediate the oxidation of hydrogen. These enzymes produce electrons via hydrogen oxidation and the produced electrons are transported to cytoplasm where sulfate is reduced to sulfide.

Nitrogenases

Nitrogenase is the enzyme which catalyzes hydrogen evolution. The active site if nitrogenase enzyme is entirely different from hydrogenase enzyme. This enzyme plays a major role in hydrogen synthesis of heterocystous cyanobacteria and photofermentation of photosynthetic bacteria. In the catalytic reaction of this enzyme, for every one mole of nitrogen fixation, 1 mol of hydrogen is produced. This enzyme mediated catalytic reaction requires additional energy in the form of ATP. Generally, four ATPs are required to generate one mol of hydrogen during this catalytic reaction (Eq. 7.23).

$$4ATP + 2e^- + 2H^+ \longrightarrow 2H_2 + 4ADP \qquad (7.23)$$

7.4 Strategies to Enhance Biofuel Production

7.4.1 Enzyme and Surfactant

Microorganisms are capable of secreting extracellular enzymes, which can carry out substrate hydrolysis. The substrate hydrolysis can be enhanced by bioagumenting these enzyme secreting microorganisms or by adding hydrolytic enzyme directly into the substrate. The direct addition of enzymes offers several advantages over the use of enzyme secreting microorganisms. Unlike microorganisms, enzymes remain unaffected by environmental factors such as pH and temperature. The presence of predators in the substrate hampers the hydrolytic potential of microorganisms, whereas it remains unaffected when enzymes are used. In addition, enzymes are water-soluble and can diffuse into the substrate and reach the unreachable area of the fermenter. A number of bacteria, actinomycetes, and filamentous fungi generate extracellular hydrolytic enzymes if cultured on certain substrates such as cellulose/hemicellulose. The most sophisticated biotechnological approach focuses on genetically altered bacteria that produce more enzymes. In biofuel research, the search for novel species capable of manufacturing hydrolytic enzymes has almost progressed. Nonetheless, Trichoderma sp extracellular enzyme system has been thoroughly studied along with economic development in enzyme synthesis inside the same organism. The production of cellulase by various bacteria may serve as a stepping stone that is used in polysaccharide hydrolysis. A great deal of research was initiated and proceeded in the past few decades on the sustainable application of agro-industrial biofuels in most industries, such as coconut oil cake, apple pomace, cassava wastewater, ground nutshell waste, grape pomace, rice straw, wheat straw, sugarcane bagasse, wheat bran, coir pith, bamboo, banana pseudo stem and so on, to generate cellulase and xylanase by fungal species such as Aspergill to generate enzymes during both solid state (SSF) and submerged (SmF) fermentation.

Surfactants are the chemical substances that induce aqueous solubilization to greater extent. Thus hasten the speed of solubilization of solid substances to the aqueous region, set free the trapped enzymes and also divulge more substrate. By taking out the EPS with surfactant, the biopolymers (proteins and carbohydrates) can be liquefied and then dissolute into the liquid part, as surfactant posses the characteristics of dissolution (Myers 2006). Recently, biosurfactants have attracted more attention due to their low toxicity, better environmental compatibility, biodegradability, and effectiveness in a wide variety of pH and temperatures.

7.4.2 Co Digestion

Numerous researches have demonstrated that lignocellulosic biomass co-digestion with a range of different co-substrates can boost methane output. When two or more feeds with opposing properties are co-digested, a favorable carbon to nitrogen (C/N) ratio of 20–30 is achieved, as well as enhanced nutritional content, the ability to neutralize inhibitory or poisonous chemicals, and improved buffering capacity. Moreover, co-digesting several industrial effluents produced in the same facility (e.g. Agri-residues and animal manure) is less cost-effective than using separate waste treatment techniques for each waste stream [38]. Lignocellulosic substrates have been employed in co-digestion with nitrogen-rich substrates like animal dung because of their high C/N ratio.

Although adopting certain waste streams to increase hydrolysis rates might be advantageous were few investigations have explored that the co-digestion is efficient. Synergisms formed by the introduction of a co-substrate with complementary qualities have enhanced the hydrolysis rate or feedstock biodegradability. To assess such effects, short-term BMP tests are often used [39]. Continuous and longer-term investigations are recommended since an inoculum that has not yet acclimated to the features of the new co-substrate will not be able to predict synergistic or antagonistic effects properly. Before using a co-substrate in a full-scale process, it's critical to define and evaluate it. Wastes that appear to be ideal co-substrate may have the opposite impact. During BMP studies to assess co-digestion of various waste combinations, both beneficial and detrimental effects were detected (slaughterhouse waste, manure, municipal solid waste and various crops). In the subsequent consistent reactor analysis, the same researchers linked synergistic effects during co-digestion to a proportionate quantity of fats, proteins, and carbs, as opposed to extremely nutritious monodigestion of slaughterhouse waste. The methanogenic community was unaffected by co-digestion, however mono-digestion of slaughterhouse waste resulted in the production of large quantities of long chain fatty acids, which impeded the system. For one of the combinations with a balanced ratio of lipids/carbohydrates/proteins, the same investigation demonstrates inhibitory activity and a drop in the relative abundance of Archaea. A substantial concentration of VFA was found in this mixture of abattoir waste and numerous crops (15 g L^{-1}). Rumen bacteria with high hydrolytic activity disrupted the balance between hydrolysis and methanogenesis, resulting in the observed negative consequences. Because the bacterial community structure was not explored, this hypothesis could not be confirmed. Only chemical and physical parameters are currently used to determine co-substrates.

Conversely, when an ultimate aim is only to maximize hydrolysis, microbiological aspects must also be examined. The rumen includes the bacteria and enzymes necessary for the decomposition of lignocellulosic material. As a result, co-digestion with rumen material might be a feasible technique for boosting hydrolysis when rumen content is accessible (for example, in a slaughterhouse waste stream). To improve hydrolysis, Napier grass and rumen content were co-digested. Cow dung was employed as a third co-substrate to deliver alkalinity due to the lower buffering ability of rumen material. Since the system's buffering capacity was optimum (pH = 7.4 and bicarbonate alkalinity 7.7 g $CaCO_3$ L-1), the amount of VS eliminated was negligible (30%). Rumen bacteria were only kept in the system when the pH was low, according to 16S rRNA gene sequencing data. As a result, a two-phase technique employing rumen material as a co-substrate in the first phase and cow dung in the second phase may have enhanced the results [40].

7.4.3 Bioaugmentation

Bioaugmentation refers to manual addition of specialized bacterial culture to the fermenter, with the intention of improving the process efficiency. Such kind of environment will be containing microbial population already; foundation resides to enrich this population and benefiting it to become more effective in converting the substrate. Adding pre-grown microbial cultures to bioreactors and other treatment equipment increases its optimal levels of efficiency without any delay. Boosting maize stover methane production during AD has also been done with a pure culture of the fermenting bacterium Acetobacteroides hydrogenigenes. *Piromyces rhizinflata YM600*, an anaerobic rumen fungus, was used to bioaugment the hydrolytic reactor of a two-phase AD system digesting cattail and corn silage, in addition to anaerobic bacteria. This method increased methane output and digesting time for a short time before the fungi's growing conditions became unfavorable [41].

Bioaugmentation cultures have been developed to continually breakdown a certain substrate that would be the most energy-intensive. Immobilization of microbial biomass on a high specific surface area adherence substrate might be a way to keep microbial populations that boost hydrolysis rates (e.g. zeolites). Enriched hemicellulolytic bacteria were immobilised on trace metal activated zeolite (aluminosilicates with high adsorption capacity) in a batch anaerobic digester and increased xylanase activities and biogas output by 162% and 53%, respectively, when compared to the control group. Zeolites can function as microbe transporter in both batch and continuous bioreactors.

Substrate modification

Intracellular lipids is the most useful biopolymer for fermentation process. On the other hand at higher concentration it resist to undergo fermentation and biodegradation. Eventhough lipids have a high biofuels production potential when its concentration exceed 30%, it happen to inhibit fermentation. During fermentation process, acidogenic microbes secrete lipid-digesting enzymes for example lipases, which digested it into glycerol and free LCFA. Glycerol is further converted into acetate by metabolic activity of acidogenesis.

Similarly, LCFA is converted into acetate by acidogenes and hydrogen by syntrophic acetogens via beta oxidation pathway. The hydrogen gas produced during LCFA oxidation was consumed by hydrogen consuming methanogens. To reduce the complexity of lipid hindrance during AD process. Use of lipid extracted biomass can be used. The extracted lipid can be used to produce biodiesel [42]. The highest biofuels yield of 236 mL CH_4/g VS was achieved by Hernández et al. [43] using lipid extracted MB. The extracted yielded 27 mg biodiesel/g biomass.

7.5 Integrated Systems for Biofuel Production

Biofuel via mono process has some limitations. In case of light dependent hydrogen generation, the production rate is very low and it is not viable due to lesser light conversion efficiency. It is not possible to achieve hydrogen yield in excess of 4 mol H_2 per mol of hexose and it is only 33% of the stochiometric conversion of glucose to hydrogen. After fermentation process, the remaining volatile fatty acids rich effluent can barely be utilized and cause dissipation of energy. Various integrated approaches are developed to make biohydrogen production feasible. Various integrated approaches are discussed below in detail.

Integrated dark fermentation and methane production system

This is an integrated system combining dark fermentative hydrogen production with that of methane production process. In this system, dark fermentation and methanogenesis are carried out in separate bioreactors. The effluent from dark fermentation hydrogen reactors are utilized as substrate for second methane production reactor. Both hydrogen and methane reactors are run at optimal conditions. The main advantage of this integrated system is enhanced waste biodegradation and biofuel production rate.

Integrated dark and photo fermentation system

This is an integrated two stage approach coupling dark fermentation as first stage and photofermentation as second stage in the reactor. In this approach, the liquid effluents from first stage reactor are utilized as substrate for second stage reactor. This combination of dark and photo fermentative process can attain theoretically a yield 12 mol of hydrogen from glucose. The reaction was represented below:

$$C_6H_{12}O_6 + 6H_2O \rightarrow 12H_2 + 6CO_2 \text{ [29]}$$

This integrated system is an effective approach to enhance biohydrogen production, increase energy recovery. This dark photo fermentation can exist as either combinative process or dual phase sequential reaction for hydrogen generation. The lab scale application of this combined system yet has many limitations to be solved. As different microbial populations are employed in in both the reactors, the microbial biomass in the effluent of the first reactor has to be separated or the effluent has to be sterilized before feeding into the second reactor. In addition, the effluent of the first reactor requires some pretreatments or preprocessing such as adjustment of pH, removal of sulfur and nitrogen. Besides, if complicated organic wastes are utilized as feedstocks in dark fermentation, then the effluent generated from dark fermentation is dark in colour and consist of huge quantity of suspended matter. These suspended materials in turn impact the photo conversion potential of microbes in photofermentative reactor.

Integrated dark fermentation and microbial electrochemical system

In this integrated system, dark fermentation process is combined with microbial electrochemical system (MECs) in which dark fermentative effluent from first hydrogen reactor is utilized as substrate in MECs system for hydrogen production. An additional voltage is applied in the MECs. Theoretically, this process can involve complete conversion of hexose and achieve a higher yield of 12 mol hydrogen/mol of hexose with an input voltage of 0.14 V. The main advantages of this integrated system are they can be employed to treat diverse variety of substrates for hydrogen production, enhanced hydrogen yield and rate. This system also has few limitations such as MECs are pH sensitive. Therefore, the dark fermentative effluent from dark fermentation reactor has to be buffered prior to electrolysis which is an expensive procedure. In addition, effective system regulation and low-cost electrode materials are essential in MECs.

Integrated dark fermentation and Microbial fuel cell

In integrated dark fermentation and microbial fuel cell (MFC) system, the fermentative effluent from dark fermentation is utilized in MFC and converted into electrical power. In this system, the particulates can be hydrolyzed and acidified in dark fermentation reactor and produces hydrogen. The effluent from this reactor is utilized in MFCs and enhances its operation. In MFCs, the hydrolyzed organic matters are oxidized by the microbes. This induces the generation of electrons. These electrons are transferred from anode to cathode via an external circuit. The electrical power is generated from electron flow. Then the protons are transferred from anode to cathode via proton exchange membrane. These protons react with oxygen and generates water.

7.6 Application of Genetic Engineering in Biofuel Production

Many approaches are employed to genetically or metabolically engineer the microbes or biocatalyst and pathways involved in biofuel production. These approaches includes genetic modification of enzyme involved in metabolic reaction (deletion and insertion of enzyme, increasing the oxygen tolerant level of enzyme) during biofuel production, modification of metabolic pathways (to insert pathways to utilize wide range of substrates, to boost the biofuel production, to convert extensive range of monomers to metabolites, to insert pathways to boost biofuel production, metabolic engineering of native biofuel production pathways.

Deletion and insertion of enzymes involved in biofuel production

During biohydrogen production, the enzyme which mediates the reoxidation of produced hydrogen and presence of this enzyme is the main issue to obtain adequate amount of hydrogen. Deletion of gene encoding this enzyme can increase the production rate and yield of biofuel (hydrogen). Gene of interest of some enzyme (essential to increase biofuel production) present in the native bacteria can be expressed in other foreign bacteria to enhance the biofuel production. Example: the gene encoding the enzyme, [FeFe]-hydrogenase obtained from anaerobic bacterium, *Clostridium Pasteurianum* is over expressed in the bacterium *cyanobacterium Synechococcus sp.* to study the possibility for boosting biofuel production potential. The constructed mutant *Cyanobacterium Synechococcus sp* resulted in significant increase of biohydrogen production. Oxygen tolerance is essential in biofuel producing enzyme especially, hydrogen producing enzymes. For this gene transfer techniques are employed. The gene encoding oxygen tolerant enzymeNiFe-hydrogenase is isolated from the bacterium *Thiocapsaroseopersicina* and inserted into Cyanobacteria to improve the biofuel (biohydrogen production).

Metabolic engineering of pathways to enhance biofuel production

The metabolic engineering approaches allows methodical and quantitative investigation of metabolic pathways and uses molecular and genetic tools to alter the metabolic routes inorder to enhance the biofuel production. This approach can be employed to eliminate the restraining factors of biofuel production via increasing the electron flow to biofuel production pathways, and improving utilization of substrates. The scheme of biofuel production that depends on fixed carbon feedstocks can be categorized into 2 levels. One is acquiring and converting complicated substrates into key metabolites and second is converting metabolites into biofuel. In this scheme, metabolic engineering plays a key role in various ways:

(a) Insertion of pathways to a microbe, allowing it to utilize a diverse variety of complicated substrates directly
(b) Insertion of pathways allowing the conversion of extensive range of monomers to metabolites
(c) Increasing the biofuel naturally that are synthesized by native microbes
(d) Addition of pathways leading to novel biofuel production

References

1. Ge H, Jensen PD, Batstone DJ (2011) Increased temperature in the thermophilic stage in temperature phased anaerobic digestion (TPAD) improves degradability of waste activated sludge. J Hazard Mater 187:355–361
2. Feng L, Yan Y, Chen Y (2009) Kinetic analysis of waste activated sludge hydrolysis and short-chain fatty acids production at pH 10. J Environ Sci (China) 21:589–594
3. Veeken A, Hamelers B (1999) Effect of temperature on hydrolysis rates of selected biowaste components. Bioresour Technol 69:249–254
4. Bolzonella D, Fatone F, Pavan P, Cecchi F (2005) Anaerobic fermentation of organic municipal solid wastes for the production of soluble organic compounds. Ind Eng Chem Res 44:3412–3418
5. Cirne DG, Agbor VB, Björnsson L (2008) Enhanced solubilisation of the residual fraction of municipal solid waste. Water Sci Technol 57:995–1000
6. Sesay ML, Ozcengiz G, Dilek Sanin F (2006) Enzymatic extraction of activated sludge extracellular polymers and implications on bioflocculation. Water Res 40:1359–1366
7. Song Y-C, Kim M, Shon H, Jegatheesan V, Kim S (2018) Modeling methane production in anaerobic forward osmosis bioreactor using a modified anaerobic digestion model No. 1. Bioresour Technol 264:211–218
8. Vane LM (2005) A review of pervaporation for product recovery from biomass fermentation processes. J Chem Technol Biotechnol 80:603–629
9. Vea EB, Romeo D, Thomsen M (2018) Biowaste valorisation in a future circular bioeconomy. Procedia CIRP 69:591–596
10. Wang L, Agyemang SA, Amini H, Shahbazi A (2015) Mathematical modeling of production and biorefinery of energy crops. Renew Sustain Energy Rev 43:530–554
11. Zhao X, Li L, Wu D, Xiao T, Ma Y, Peng X (2019) Modified Anaerobic Digestion Model No. 1 for modeling methane production from food waste in batch and semi-continuous anaerobic digestions. Bioresour Technol 271:109–117
12. Atkinson B, Mavituna F (1991) Biochemical engineering and biotechnology handbook, 2nd edn. Stockton Press, New York, NY
13. Singh S, Chakravarty I, Pandey KD, Kundu S (2018) Development of a process model for simultaneous saccharification and fermentation (SSF) of algal starch to third-generation bioethanol. Biofuels 1–9
14. Gregg DJ, Saddler JN (1996) Factors affecting cellulose hydrolysis and the potential of enzyme recycle to enhance the efficiency of an integrated wood to ethanol process. Biotechnol Bioeng 51:375–383
15. Gusakov AV, Sinitsyn AP (1992) A theoretical analysis of cellulase product inhibition effect of cellulase binding constant, enzyme substrate ratio, and beta-glucosidase activity on the inhibition pattern. Biotechnol Bioeng 40:663–671

16. Kadam KL, Rydholm EC, Mcmillan JD (2004) Development and validation of a kinetic model for enzymatic saccharification of lignocellulosic biomass. Biotechnol Prog 20:698–705

17. Ahlert S, Zimmermann R, Ebling J, König H (2016) Analysis of propionate-degrading consortia from agricultural biogas plants. Microbiologyopen 5:1027–1037

18. Wu H, Fu Y, Guo C, Li Y, Jiang N, Yin C (2018) Electricity generation and removal performance of a microbial fuel cell using sulfonated poly (ether ether ketone) as proton exchange membrane to treat phenol/acetone wastewater. Bioresour Technol 260:130–134

19. Garrote G, Domínguez H, Parajó JC (1999) Hydrothermal processing of lignocellulosic materials. Eur J Wood Wood Prod 57:191–202

20. Mao C, Feng Y, Wang X, Ren G (2015) Review on research achievements of biogas from anaerobic digestion. Renew Sustain Energy Rev 45:540–555

21. Tsui T-H, Chen L, Hao T, Chen G-H (2016) A super high-rate sulfidogenic system for saline sewage treatment. Water Res 104:147–155

22. Shuai L, Luterbacher J (2016) Organic solvent effects in biomass conversion reactions. Chemsuschem 9:133–155

23. Nitsos CK, Matis KA, Triantafyllidis KS (2013) Optimization of hydrothermal pretreatment of lignocellulosic biomass in the bioethanol production process. Chemsuschem 6:110–122

24. Alonso DM, Bond JQ, Dumesic JA (2010) Catalytic conversion of biomass to biofuels. Green Chem 12:1493

25. Li R, Xie Y, Yang T, Li B, Wang W, Kai X (2015) Effects of chemical-biological pretreatment of corn stalks on the bio-oils produced by hydrothermal liquefaction. Energy Convers Manag 93:23–30

26. Holzhäuser FJ, Creusen G, Moos G, Dahmen M, König A, Artz J, Palkovits S, Palkovits R (2019) Electrochemical cross-coupling of biogenic di-acids for sustainable fuel production. Green Chem 21:2334–2344

27. Savage PE (2009) A perspective on catalysis in sub- and supercritical water. J Supercrit Fluids 47:407–414

28. Bedoić R, Čuček L, Ćosić B, Krajnc D, Smoljanić G, Kravanja Z, Ljubas D, Pukšec T, Duić N (2019) Green biomass to biogas—a study on anaerobic digestion of residue grass. J Clean Prod 213:700–709

29. Biernacki P, Steinigeweg S, Borchert A, Uhlenhut F (2013) Application of Anaerobic Digestion Model No. 1 for describing anaerobic digestion of grass, maize, green weed silage, and industrial glycerine. Bioresour Technol 127:188–194

30. García-Diéguez C, Bernard O, Roca E (2013) Reducing the Anaerobic Digestion Model No. 1 for its application to an industrial wastewater treatment plant treating winery effluent wastewater. Bioresour Technol 132:244–253

31. Ntaikou I, Gavala HN, Lyberatos G (2010) Application of a modified Anaerobic Digestion Model 1 version for fermentative hydrogen production from sweet sorghum extract by Ruminococcus albus. Int J Hydrogen Energy 35:3423–3432

32. Pastor-Poquet V, Papirio S, Steyer J-P, Trably E, Escudié R, Esposito G (2018) High-solids anaerobic digestion model for homogenized reactors. Water Res 142:501–511

33. SriBala G, Carstensen H-H, Van Geem KM, Marin GB (2019) Measuring biomass fast pyrolysis kinetics: state of the art. Wiley Interdiscip Rev Energy Environ 8:e326

34. Gama FM, Teixeira JA, Mota M (1994) Cellulose morphology and enzymatic reactivity: a modified solute exclusion technique. Biotechnol Bioeng 43:381–387

35. Klyosov AA (1990) Trends in biochemistry and enzymology of cellulose degradation. Biochemistry 29:10577–10585

36. Kristensen JB, Borjesson J, Bruun M, Tjerneld F, Jorgensen H (2007) Use of surface active additives in enzymatic hydrolysis of wheat straw lignocellulose. Enzyme Microb Technol 40:888–895

37. Lenz J, Esterbauer H, Sattler W, Schurz J, Wrentschur E (1990) Changes of structure and morphology of regenerated cellulose caused by acid and enzymatic hydrolysis. J Appl Polym Sci 41:1315–1326

38. Kilpeläinen I, Xie H, King A, Granstrom M, Heikkinen S, Argyropoulos DS (2007) Dissolution of wood in ionic liquids. J Agric Food Chem 55:9142–9148
39. Brandt A, Hallett JP, Leak DJ, Murphy RJ, Welton T (2010) The effect of the ionic liquid anion in the pretreatment of pine wood chips. Green Chem 12:672
40. Brandt A, Gräsvik J, Hallett JP, Welton T (2013) Deconstruction of lignocellulosic biomass with ionic liquids. Green Chem 15:550
41. Wahlström RM, Suurnäkki A (2015) Enzymatic hydrolysis of lignocellulosic polysaccharides in the presence of ionic liquids. Green Chem 17:694–714
42. Chen Q, Liu D, Wu C, Xu A, Xia W, Wang Z, Wen F, Yu D (2017) Influence of a facile pretreatment process on lipid extraction from Nannochloropsis sp. through an enzymatic hydrolysis reaction. RSC Adv 7(84):53270–53277
43. Hernández D, Solana M, Riaño B, García-González MC, Bertucco A (2014) Biofuels from microalgae: lipid extraction and methane production from the residual biomass in a biorefinery approach. Biores Technol 170:370–378

Chapter 8
Single Stage Anaerobic Digestion

8.1 Introduction

Anaerobic digestion involves a complex consortium of microorganisms. The reaction scheme for anaerobic digestion is presented. Six distinct processes identified in the anaerobic digestion of biopolymers (carbohydrates, proteins, and lipids) are:

i. Hydrolysis of biopolymers

    i. Hydrolysis of Proteins
    ii. Hydrolysis of Carbohydrates
    iii. Hydrolysis of Lipids

ii. Fermentation of amino acids and sugars
iii. Anaerobic oxidation of long-chain fatty acids
iv. Anaerobic oxidation of intermediary products like volatile fatty acids
v. Conversion of acetate to methane
vi. Conversion of hydrogen to methane.

    The metabolism of the various bacterial groups is interdependent, and all the different phases in metabolism cannot be clearly delineated. van Handel and Lettinga state that only the following phases could be distinguished clearly [1].

(i) Hydrolysis

In this phase, organic polymers (carbohydrates, proteins, lipids, and organic matter) are broken down into dissolved compounds with lower molecular weight. Carbohydrates are transformed into soluble sugars; proteins are degraded to amino acids and lipids are converted to long chain fatty acids and glycerin by fermentative bacteria.

(ii) Acidogenesis

The end products of hydrolysis are fermented into organic acids, volatile fatty acids and hydrogen, carbon dioxide, ammonia, and hydrogen suphide by obligate anaerobic bacteria that grow well in pH between 5 and 6.

© Springer Nature Singapore Pte Ltd. 2022
K. Sudalyandi and R. Jeyakumar, *Biofuel Production Using Anaerobic Digestion*,
Green Energy and Technology, https://doi.org/10.1007/978-981-19-3743-9_8

(iii) Acetogenesis

The products of acidogenesis viz., hydrogen, carbon dioxide, and acetate are converted into acetic acid accompanied by the formation of more hydrogen and carbon dioxide. More than 70% of the initial COD of the influent will be converted into acetic acid and the rest into hydrogen.

(iv) Methanogenesis

The energy (Methane) Producing phase of anaerobic treatment is called methanogeneis. Methanogenesis is an important terminal electron accepting process in many anaerobic environments where the supply of oxygen, nitrate, and oxidized forms of sulphur, iron, and manganese are limited (Benefield and Randall 1980). The amount of energy released during methanogenesis is relatively low compared to other terminal electron accepting processes. Thus, the amount of biomass produced per unit of substrate degraded is much less than that of other terminal electron accepting processes. Methanogenesis is used as the treatment of choice for sewage and other complex wastes since sludge yields are low and most of the energy in the original substrates is retained in the energy rich fuel, methane [2]. The formation of methane, the ultimate product of anaerobic digestion, occurs through two major routes. The primary route is through fermentation of acetic acid into CH_4 and CO_2 by acetoclastic bacteria like Methanosarcina spp. and Methanosaeta spp.

$$CH_3COOH \rightarrow CH_4 + CO_2 \qquad (8.1)$$

Hydrogenotrophic methanogenesis utilizes hydrogen to reduce carbon dioxide to methane.

$$4H_2 + CO_2 \rightarrow CH_4 + 2H_2O \qquad (8.2)$$

The bacteria producing methane from hydrogen and carbon dioxide grow faster than those utilizing acetate. There is a synergistic relation between hydrogen producers and hydrogen scavengers. At higher partial pressures of hydrogen, hydrogen oxidation is more thermodynamically favorable than acetate degradation and acetate production increases. The optimal pH for methanogens is around 7.0. Acetoclastic methanogenesis is generally the rate-limiting step in anaerobic digestion.

Based on the physical separation phases, the AD process can be divided into single stage (1S-AD) and two-stage digestion (2S-AD). The microbes involved in the AD process were categorized into acidogens and methanogens. These microbes can survive under varying ecological conditions based on which AD could be configured into 1S and 2S-AD. The digesters employed for single and two-stage AD can afford a perfect environment for diverse microbial populations and permit for healthier process regulation [3]. According to literature, two-stage was superior to single-stage AD pertaining to energy recovery and stable digestion. Only in a few pieces of literature the performance of single and two-stage AD were compared.

For instance, a comparison investigation was done by Voelklein et al. [4] to assess the influence of organic loading rates on 1S and 2S-AD of food waste. Similarly, in another study, De Gioannis et al. [5] evaluated the energy production of mesophilic 1S and 2S-AD. Some studies have suggested that thermophilic two-stage digestion was considered to be beneficial for effective energy recovery and complete substrate biodegradation. Lee et al. [6] have reported temperature phased 2S-AD benefited by greater production of hydrogen due to recirculation of methanogenic effluents into the first stage. Few reports have evaluated 1S and 2S temperature phased AD and reported inconsistent outcomes. According to Leite et al. [7] temperature phased AD increases substrate degradation and energy recovery efficiency than single-stage AD. Conversely, Schievano et al. [8] reported no considerable variation in the bioenergy yield of both temperature phased 1S and 2S-AD treating market organic waste and swine manure. Similarly, Park et al. [9] reported methane recovery in single and temperature phased two-stage AD of kitchen garbage were the same.

The collation of the stability of 1S-AD and 2S-AD was demonstrated voluminously, and merits/demerits of both the processes have been excogitated and figured out through copious researchers [10–13]. Along these lines, this paper proffers to juxtapose between 1S- and 2S-AD and contextualized hinged upon their performance, reaction processes and energy assessment in a expatiate manner.

8.2   Single Stage Digestion (1S-AD)

AD is a series of biological pathways, that involves hydrolytic, acidification, acetogenic, and biomethanation in which acetate, hydrogen, and carbon dioxide are transformed to biomethane. In single stage digestion, the entire biochemical reactions happen in one digester. Single stage digesters are known for their simplicity in which the entire biochemical processes occur in single reactor vessel. Ren et al. [14] have reported that 95% of full scale digesters treating biowastes such as biodegradable portions of municipal solid waste and biowaste in Europe are single-phase anaerobic digesters. In these types of digesters, the biochemical processes take place under the same operational conditions. These digesters have mixing and heating devices to maintain uniform operational conditions. These digesters are run under uniform feeding rate. Therefore, the volume of the reactor vessel can be compact, and the reactor performance is enhanced.

8.2.1   Digester Types

The single stage digesters are mainly classified into two types: single stage low solids or wet digestion system and single stage high solids or dry digestion system on the basis of feed materials total solids concentration [15]. Examples of single stage wet digesters are continuously stirred tank reactors (CSTR), upflow anaerobic sludge blanket reactor (UASB), upflow anaerobic solid-state reactor (UASS), anaerobic fluidized bed reactor (AFBR), upflow anaerobic filter process reactors (UAFR), expanded granular sludge bed reactors (EGSBR). Plug flow reactor (PFR) is the best example of single stage dry digesters. Table 8.1 represents the various types of single stage digesters, its operating conditions and its productivity.

Single stage low solids or wet digestion system

In single stage low solids digesters, the feasible feed solid content should be 15–25% [16]. It is generally employed for sludge reduction in wastewater treatment plants over decades. The advantage of single-stage low solids system is its ease in operation. This system has been come into operation before the development of high solids single stage digesters. Capital investments for these digesters are cheaper. However, it demands higher operational cost for internal mixing, essential dewatering and disintegration processes. The various types of single stage low solids digesters are described in following section.

Continuously stirred tank reactors (CSTR)

Continuously stirred tank reactor (CSTR) is one of the most employed low solids wet digestion reactor and can be work under both mesophilic and thermophilic condition. In these types of reactors, the introduced substrate is stirred incessantly, and the reactor contents are mixed entirely so that the accessibility of substrates to the microbial inoculum will be enhanced. Based on the substrate type, the retention period of these reactors varies. These reactors are mainly employed for treating substrate with a total solid of about 2–10%. These bioreactors are mainly employed for treating feedstocks which includes, cattle slurry, municipal solid waste, sludge, agro wastes, food waste, petrochemical wastewater [59–64] etc. The main disadvantage of these reactors is its high energy consumption and short circuiting [65]. It reduces biogas productivity of the reactors. To overcome this issue, it is essential to sterilize the substrate prior to digestion. The reactor is usually a circular closed vessel (Fig. 8.1a). Influent was fed into CSTR by means of exterior port and heat is maintained by means of heat exchangers. The substrate materials are mixed with the help of a mechanical mixer. A stiff port traps the biogas. The important operational parameters of CSTR are temperature, organic loading rate (OLR), Hydraulic retention time (HRT) etc. Nagao et al. [66] have suggested that it is possible to run single stage wet CSTR (mesophilic temperature of 37 °C) at a substrate load of 15 g COD/L/day and retention time of 16 days for treating waste activated sludge (WAS). They achieved a methane production of about 455 mL/g VS, respectively. Zarkadas et al. [63] have operated CSTR in thermophilic condition (temperature 55 °C) at an OLR and HRT of 6.91 g VS/L/d and 21 days. They have achieved a biomethane production of 385 mL/g VS respectively. Working on mesophilic AD of agro waste—rapeseed straw in CSTR, Fu and Hu [62] have achieved a methane production of 160 mL/g VS at an OLR and HRT of 3 g VS/L/d and 30 days. While co-digesting kitchen and fruits and vegetable waste (FVW) in pilot scale CSTR at an OLR and HRT of 4.5/g VS/L/d and 30 days in mesophilic temperature of 35 °C, Wang et al. [67] have achieved methane yield and percentage was 0.46 L/g VS and 64.8%. Sevilla-Espinosa et al. [68] have used CSTR

Table 8.1 Performance of various single stage digesters and its operating conditions

S. No	Digesters	Substrate used	Temp (°C)	OLR	HRT (days)	VS removal (%)	Methane yield (or) productivity	Methane (%)	References
1	CSTR	Food waste + cattle slurry	36	2 g VS/L/d	30	NR	0.306 L/g VS	62.7	[17]
2	CSTR	FPW (Food processing industry wastes)	55	3 g VS/L/d	7	94	0.44 L/g VS	84	[18]
3	CSTR	OFMSW	55	3 g VS/L/d	10	93	0.42 L/g VS	67	[18]
4	CSTR (Semi continuous)	WAS + Food waste	55	1 g VS/L/d	33.3	75	0.533 L/g VS	NR	[19]
5	CSTR	Rapeseed straw	37	3 g VS/L/d	30	36.6	160 mL/g VS	58	[20]
6	CSTR	Pasteurized food waste + cattle manure	55	6.91 g VS/L/d	21	72.1	385 mL/g VS	86	[21]
7	CSTR	Petrochemical wastewater	55	7.5 g COD/L/d	4	NR	0.6 L/gCOD	65.49	[22]
8	Single wet CSTR	WAS	37	15 g COD/L/day	16	92.2	455 mL/g VS	NR	[23]
9	Pilot scale CSTR	Kitchen waste + Fruits and vegetable waste	35	4.5 g VS/L/d	30	65.4	0.46 L/g VS	64.8	[24]
10	UASB	Potato juice	37	5.1 g VS/L/d	5	NR	240 mL/g VS	51	[25]
11	EGSB	Potato juice	37	3.2 g VS/L/d	8	NR	380 mL/g VS	81	[25]
12	UASB	Wheat straw stillage + pig manure	55	17.1 g COD/L/d	48 h	NR	154.8 mL/g COD	63.7	[26]

(continued)

Table 8.1 (continued)

S. No	Digesters	Substrate used	Temp (°C)	OLR	HRT (days)	VS removal (%)	Methane yield (or) productivity	Methane (%)	References
13	UASS	Corn stalk	39	2.5 g VS/L/d	18	NR	184 mL/g VS	54.4	[27]
14	UASS	Horse manure	35	4.5 g VS/L/d	7	69	86.9 L/kg VS	NR	[28]
15	PFR	Food waste	35	NR	60	NR	NR	60	[29]
16	PFR	Food waste	31	1.12 g VS/L/d	20	91.37	0.341 L/g VS	65	[29]
17	PFR	Garden and kitchen waste	35	3.1 g VS/L/d	20	NR	0.105 L/g VS	62	[30]
18	Continuous single stage digester	Food waste	50	1 g VS/L/day	28	91%	0.78 m³/kg VS	74	[31]
19	Single stage fed batch digester	Fruits and vegetables waste	28–46	NR	98	NR	0.387 m³/kg VS	65	[32]
20	Thermophilic lab scale semi continuous digester	Food wastewater + WAS	55	NR	20	73	0.316 L/ gCOD	68.24	[33]
21	Mesophilic lab scale semi continuous digester	Food wastewater + WAS	35	NR	20	71	0.268 L/ gCOD	65.21	[33]
22	CSTR	Cow manure + fruits and vegetable waste + corn straw	38	2.61 g VS/L/d	15	NR	743.24 mL/g VS	NR	[34]

(continued)

Table 8.1 (continued)

S. No	Digesters	Substrate used	Temp (°C)	OLR	HRT (days)	VS removal (%)	Methane yield (or) productivity	Methane (%)	References
23	Lab scale single stage digester	Food wastewater	35	7 g COD/L/day	20	80	0.28 L/g VS	NR	[35]
24	CSTR	Brown water + Food waste	35	0.8 g VS/L/d	40	85	NR	62	[36]
25	CSTR	Swine slurry + market biowaste	55	NR	25	NR	NR	54	[37]
26	CSTR	Coffee production waste	35	10.28 g COD/L/d	40	52	327 mL/g VS	65	[38]
27	CSTR	OFMSW	55	30.7 g VS/L/d	3	45	0.26 L/g VS	61	[39]
28	CSTR	Food waste + piggery wastewater	35	NR	20	NR	396 mL/g VS	NR	[40]
29	Single stage mesophilic reactor	Concentrated WAS	35	NR	30	37.6	0.38 L/g VS	60.6	[41]
30	Horizontal mid scale anaerobic single stage reactor	Food waste + chinese silver grass	36	2.1 g VS/L/d	NR	NR	268.4 mL/g VS	60	[41]
31	Single stage mesophilic anaerobic digester	Sewage sludge	35	1.43 g VS/L/day	20	43.5	451 mL/g VS	64.7	[42]

(continued)

Table 8.1 (continued)

S. No	Digesters	Substrate used	Temp (°C)	OLR	HRT (days)	VS removal (%)	Methane yield (or) productivity	Methane (%)	References
32	Single stage thermophilic anaerobic digester	Sewage sludge	55	2.90 g. VS/L/day	10	46.8	416 mL/g VS	63.6	[43]
33	CSTR	Turkey manure	37	1 g VS/L/day	160	87.6	NR	70	[44]
34	CSTR	Municipal biomass + activated sludge	35	8 g VS/L/d	15	69.9	0.62 L/g VS	55.6	[45]
35	UASB	Food waste	35	9.1 g SCOD/L/d (or) 7.05 g VS/L/d	2.2	61	0.0.35 L/g SCOD (or) 0.452 L/g VS	75	[45]
36	Mesophilic completely mixed anaerobic digester	Food waste + dairy manure	36	3 g VS/L/d	NR	82.7	0.51 L/g VS	72	[46]
37	Lab scale semi continuous digester	Food waste + cattle manure	35	15 g VS/L/d	NR	NR	317 Ml/Gvs	60.2	[47]
38	Anaerobic fluidized bed membrane bioreactor	Synthetic wastewater	25	2 g COD/L/d	3 h	100	NR	53	[48]

(continued)

Table 8.1 (continued)

S. No	Digesters	Substrate used	Temp (°C)	OLR	HRT (days)	VS removal (%)	Methane yield (or productivity)	Methane (%)	References
39	Lab scale semi continuous digester	Synthetic medium-methanol rich wastewater	55	12 g COD/L/d	30	NR	140 mL/g COD	NR	[49]
40	Lab scale batch single stage digester	Lipid extracted microalgal biomass	35	2 g VS/L/d	15	NR	195 mL/g VS	NR	[50]
41	CSTR	Acidified cattle manure + normal cattle manure	50	NR	20	33	203.8 mL/ g VS	NR	[51]
42	UASB	Coffee wet wastewater	35	3.6 g COD/L/d	21.5	NR	NR	58	[52]
43	AFFBR	Synthetic wastewater	25	1.387 g COD/L/d	36 h	NR	0.44 L/g COD	64.8	[53]
44	AFFFBR	Synthetic wastewater	25	1.387 g COD/L/d	36 h	NR	0.42 L/g COD	61.5	[53]
45	EGSBR	Cheese whey	25	7.5 g COD/L/d	6	NR	340 mL/g COD	63.5	[54]
46	EGSBR	Vinasse	25	5.8 g COD/L/d	6	NR	245 mL/g COD	70.8	[54]
47	EGSBR	Coffee processing wastewater	25	3 g COD/L/d	6	NR	300 mL/g COD	80.3	[54]

(continued)

Table 8.1 (continued)

S. No	Digesters	Substrate used	Temp (°C)	OLR	HRT (days)	VS removal (%)	Methane yield (or) productivity	Methane (%)	References
48	EGSBR	Coal gasification wastewater	35	625.25 mg COD/L/d	48 h	NR	NR	60	[55]
49	UASB	Wheat straw + seaweed hydrolysate	37 lePara>	1.2 g COD/L/d	7.5	NR	0.3 L/g COD	55	[56]
50	UASB	Vinasse of banana waste	30	17.05 g COD/L/d	7.5	NR	0.263 L/g COD	84	[57]
51	CSTR	Dairy manure + maize silage	37	1.1 g VS/L/d	30	67	425 mL/g VS	60	[58]

for treating substrate such as cheese whey, vinasse and coffee processing wastewater. During reactor operation, they have fixed HRT (6 days) and optimized OLR as 7.5, 5.8 and 3.

g COD/L/d for treating cheese whey, vinasse and coffee processing wastewater respectively. At their respective optimum OLR, CSTR achieved a methane yield of about 340, 245 and 300 mL/g COD. Fang et al. [69] have treated potato juice in CSTR at mesophilic temperature of 37 °C at a substrate load of 3.2 g VS/L/d and achieved a methane production of 380 mL/g VS.

Upflow anaerobic sludge blanket reactor (UASB)

Upflow anaerobic sludge blanket reactor (UASB) is the high-rate single stage wet digester mainly employed for treating wastewater. Rarely co-digestion of solids in UASB was reported in literature [70, 71] These reactors were known for its ability of self-granulation of anaerobes and were firstly discovered by Lettinga in the 80s. These reactors were attractive owing to its eminent feed rate, reduced retention period,

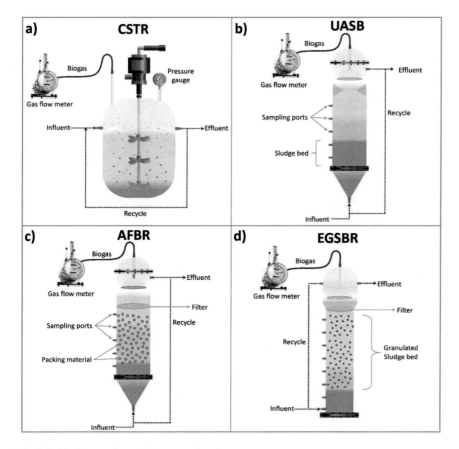

Fig. 8.1 Single stage low solids or wet digestion system

minimal operating cost, reduced sludge generation and efficient methane yield [72, 73]. In these digesters (Fig. 8.1b), the organic rich influent wastewater enters the digester through the bottom. The substrate wastewater is then passed via the sludge blanket (granulated biomass). In the sludge blanket the organics were digested in an anaerobic condition. A gas liquid separator in the reactor separates the generated biogas. The separated liquor is cleared above a barrier and the biogas is collected through the bottom port of the reactor. At same time, the granulated biomass remains at the bottom of digester [74]. Predominantly, the development of bacilli type bacteria such as *Methanothrix spp* that forms round shaped granules impacts the granulation process during UASB operation [75]. The parameters that effectively influence the USAB functioning includes thickness of flocs, compactness of microbial population and configuration of gas liquid separator, which efficiently enhances the retention of microbial biomass in the digester. In a lab scale UASB, the achieved methane was found to be 58% while treating coffee wet wastewater at a substrate load and retention time of 3.6 g COD/L/d and 21.5 days in a mesophilic range of temperature. An UASB at thermophilic condition could achieve methane production of about 154.8 mL/g COD at an retention time and substrate load of 48 h and 17.1 g COD/L/d, respectively [74]. For better operation of UASB, the following conditions should be taken into account: Choosing of appropriate substrate (wastewater) having the ability of self-granules formation; reactor operation without using mechanical agitators and prevention of bulking.

Upflow anaerobic solid state reactor (UASS)

Upflow anaerobic solid state reactor (UASS) in the novel type of bioreactor designed for treating various types of substrates like wheat straw, maize silage, horse manure, corn stalk, etc. [75, 77]. An impetuous solid liquid separation induced by the variation in substrate density and the reactor liquid is the principle behind these types of reactors [78]. In these types of bioreactors, the substrate is fed through an inclined pipe located at the lowest portion of the digester. The substrate is arranged as solid state bed. The flow of the substrate was upflow and therefore the digester was termed as upflow anaerobic solid state reactor [75]. In mesophilic UASS digesting wheat straw. A higher methane yield was observed [78] of 91 L/kg VS at a substrate load of 2.5 g VS/L/d and at a HRT of 14–21 days. In similar studies, Pohl et al. [79] have achieved greater biomethane production of 161 L/kg VS at thermophilic conditions [75, 80] The disadvantages of UASS was attributed to its configuration as it limits amount of substrate to be treated.

Anaerobic fluidized bed reactor (AFBR)

In these digesters, the medium for microbial adherence is fluidized inside the reactor [72]. Generally, sand or alumina is employed as the attachment medium. But lately medium with lower density such as anthracite or plastic materials are used, and it minimize the operational cost of AFBR. The medium used for microbial attachment increases the mass transfer in the reactor. The clogging problem is minimal in AFBR. The volume of the bioreactor is reduced due to increased surface area. Therefore, yield is higher in AFBR In full scale implementation, these reactors have complications in regulating the size and density of particles leading to instability in operation [81]. These reactor can be operated at high OLR, which in turn enhances the performance of the reactor [64]. In some cases, the microbes attached to the support medium grow as thick biofilm and get detached from the medium. As a result, the particles move up and washed out from the reactor leading to loss of biomass concentration in the reactor. This can be minimized by recycling the biomass again into the reactor. These reactors also have a disadvantage of high energy consumption which increases the operational cost. Figure 8.1c shows the schematic representation of AFBR [82, 83].

Upflow anaerobic filter reactors (UAFR)

Upflow anaerobic filter reactor (UAFR) is the single stage digester first developed by McCarty in 60s. It was mainly employed for treating wastewaters generated from industries and wastewater plants. These types of reactors contain a medium such as rocks and plastics for microbial attachment. The microbial population may present within the gap of the medium as well as in the surfaces of the medium. As a result, dense populations of microbes are retained in the digester. In UAFR, a distributor is fitted at the bottom to circumvent short circuiting flow via the packed column and to provide homogenous upflow of wastewater [72]. It is observed that about 90% COD removal in UAFR treating complex dairy wastewater at industrial scale [84].

Expanded granular sludge bed reactors (EGSBR)

Expanded granular sludge bed reactor (EGSBR) bear a resemblance like UASB. Figure 8.1d shows the schematic of EGSBR. Granulated sludge is maintained in EGSBR and it get expanded during greater hydraulic rates. The granular sludge is partly fluidized by liquid recirculation at an upflow velocity of 5–6 m/h. These reactors have two main compartments: digester vessel and gas- solid -liquid separator. EGSBR has been mainly used in the treatment of soluble effluents. This is because of the presence of greater hydraulic velocities within the digesters that makes the removal of particulate organics difficult. Greater mass transfer efficiency of EGSBR makes it attractive over other anaerobic treatment system. In EGSBR, the applied upward velocity causes a considerable reduction in the space. This facilitates the digestion process easier. These reactors are more suitable for full scale implementation when compared to fluidized bed reactors. The EGSBR was advantageous than UASB due to efficient mixing owing to greater upflow velocity, enhanced mass transfer efficiency, better viability of biomass, and efficient diffusion of substrate to sludge flocs. It is demonstrated that EGSBR can be operated at greater organic and hydraulic loadings for treating acid containing wastewater in psychrophilic circumstances [85]. In addition, EGSBR was proficient for biodegrading lipids, inhibitory compounds and soluble pollutants rich effluent.

8.2.2 Single Stage High Solids or Dry Digestion System

In single stage high solids or dry digesters, the solid content should be greater than 30% [16]. The major problem of these reactor is handling and transport of substrate due to its high solids content. The equipments are expensive when compared to low solids single stage digesters. Research has been focused towards dry systems in the last decades than wet systems. This is because the dry systems are found to be advantageous than wet systems in certain circumstances. For instance, disintegration process is simpler in this high solid system when compared to low solids system since it possesses greater tolerance to materials like glassy product, wood chips, etc. Therefore, these reactors are appropriate for digesting mechanically sorted OFMSW (organic fraction of municipal solid waste) and it minimizes the organic loss which in turn enhances the productivity. In addition, in these digesters coarse particles alone has to be removed prior to process of digestion, i.e. greater than 40 mm. The substrate for single stage high solid digestion system is viscous in nature [15]. Due to high viscous, the dry viscous particles cannot move freely through the digester. Therefore, heavier processing equipments are needed to run the single stage dry digester in continuous mode. The single stage high solids digester (plug flow reactor) is described beneath.

Plug flow reactors (PFR)

Plug flow reactor (PFR) is one of the high solids or dry digestion system which is mainly used for treating slurries, concentrated manure, cattle deposits, organic portion of solids waste, municipal solid waste etc. [15] and could be run at both mesophilic and thermophilic temperatures. PFR was covered with a gas tight or inflatable cover to store fuel gas. PFR cannot be employed for treating substrate with low solids concentration as it causes floating or settling of deposit. In PFR, the SRT (solids retention time) is equivalent to HRT. The organic loading rates of PFR are very low when compared to other reactors. PFR have no internal mixing and it can be operated under HRT of 15–29 days. It demands low investment, works under ease operation and results in higher production of biofuel while treating digested semi solid organics. Over the years, PFR was successfully employed by researchers to generate biogas from high solids organic wastes. The efficiency of PFR for treating food waste and reported 60% methane production at 60 days HRT under mesophilic condition [86–88].

Table 8.1 list the performance and productivity of various lab scale single stage digesters and discussed their deliverable. The productivity and methane yield may vary depending upon the operational parameter employed. It is ascertained from several studies that the thermophilic reactor performs better than mesophilic and their respective biomethane production, biomethane percentage and solids reduction were 0.316 L/g COD, 68.24 and 73% for thermophilic and 0.268 L/g COD, 65.21 and 71% for mesophilic. Accordingly it can be concluded that EGSBR and UASS are considered more efficient in terms of process efficiency, high organic loading rates, upflow velocities, higher contact between substrates and biomass, less operating and investment cost when compared to others. Among them, UASS is feasible for long term operation and devoid of process instabilities.

8.3 Microbial Communities and Pathways

Microbial population determine the biodegradation potential of substrates in AD. They are regarded as central core of the reactors. Operation of a single stage digester is significantly affected by the changes in microbial population and metabolic pathways involved [89]. The performances of AD are typically governed by the dynamic microbial population [90]. Effectiveness of a digester relies on the energetic microbial population. These microbial population through their various metabolic reactions imparts a significant task in digestion process. Microbial populations in digester are well associated with the nutrients and environmental condition of the reactor. Studying and exploring its population changes, diversity and dynamics will help to maintain the process stability in the digester. This can be done by optimizing the process parameters and by improving the favored metabolic pathways which in turn increases the productivity and yield.

8.3.1 Microbial Communities Abundance and Its Changes

AD is a multifaceted biochemical reaction including various phases: hydrolytic, acidogenic, acetogenic and methanogenic phases. These sequential biochemical reactions are mediated by wide diverse of microbial populations [91]. Analysis of microbial communities is regarded as an essential factor that regulates the biodegradation potential of substrate in AD. Microbes that mediate AD process could be categorized into two types: bacteria and archaea. Bacterial population degrade larger substances into organic acids, carbondioxide and hydrogen and archaea are accountable for methane generation. Therefore, studying the multiplicity and activity of the microbial population can help in optimization of processes in AD. Figure 8.2 shows a schematic representation of different microbial communities abundance in single-stage AD. Recent advancement in molecular practices which includes popular high-throughput sequencing, metagenomics [92], quantitative real-time polymerase chain reaction (qPCR) [93] and pyrosequencing [94], helps to detect changes in microbial communities during reactor operation and to optimize the process parameters. The qPCR facilitates real-time recognition and quantitative analysis of genome in a DNA sample. In pyrosequencing, thousand to millions of sequences could be sequenced in a single lane. Thus it facilitate the possibility of analyzing entire microbial population in AD process [95–97]. In throughput sequencing, sequences of multiple samples can be detected using specific barcode, to analyze diverse various microbial population [98].

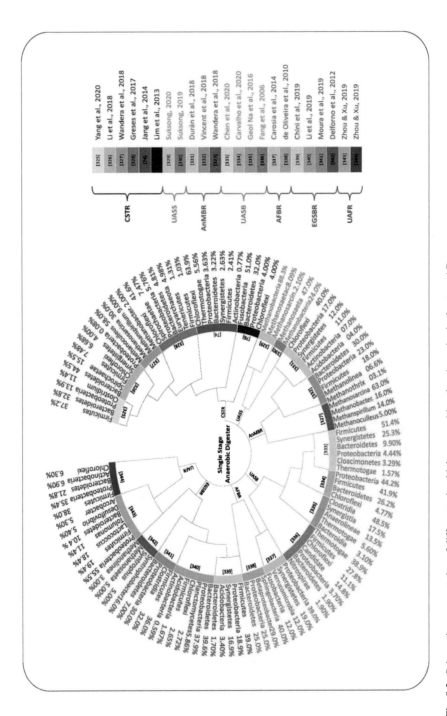

Fig. 8.2 Schematic representation of different microbial communities abundance in single-stage AD

The changes in digester operational parameters, environmental factors and nature of substrate can influence microbial population. Hence, optimizing the above can regulate microbial population and process efficiency during AD. As mentioned earlier, AD process and its performance is dictated by diverse microbial populations, which mediate sequential biochemical reactions. Any slight variations in operational parameters negative influence in composition and activity of microbial community leading into the reactor instability. On the ther hand, if any biological benefit is probable among the microbes in the digesters, the process stability of the digester will not be influenced. Under instable reactor condition, the structure and richness of microbial populations get altered [99]. In such circumstances, a detailed knowledge about microbial community succession, association between microbes, relationship between microbes and process dynamics gives a chance for a process engineer to revert the system to stable. Table 8.2 represents the summary of microbial communities abundance in various single stage digesters at varying operational conditions. In single stage digestion changes or shift in microbial population in single stage digestion during environmental or process disturbances (stress tolerance) [100, 101]. For instance, changes in community structure of microbes, from acetoclastic to hydrogenotrophic methanogens, occurs during labscale single stage AD treating high strength food wastewater [102] and an obvious upsurge of the hydrogenotrophic methanogens from 16 to 41% during the entire single stage digestion. This could be due to variations in the development and growth among acetoclastic and hydrogenotrophic methanogens; For instance, (i) shorter HRT offers favorable environment for hydrogenotrophic methanogens when compared to acetoclastic methanogens (ii) an increment in OLR causes alteration in bacterial population. These changes in bacterial population in turn affect the methanogens since they get substrates (carbon) from bacterial community [94, 102]. Futher there is a linkage between microbial characteristics and process parameters [103–105]. For example, It is observed that microbe *Methanosarcina* have high tolerance to VFA concentration when compared to *Methanosaeta* [106]. In addition to Hydrogenotrophic methanogens such as *Methanoculleus* have high tolerance to salt concentration. when OLR was increased, the abundancy of acid producing and syntrophic VFA oxidizing bacterial population are also increased. While at the same time, the occurrence of syntrophic associates (hydrogenotrophic methanogens) of acid producing microbes have decreased in quantity [92].

Among the microbial population involved in AD, four phylas are dominant (*Chloroflexi, Proteobacteria, Bacteroidetes and Firmicutes*) and their comparative loads vary with system [96, 107, 108]. It is observed that the microbial population belong to the genus *chloroflexi* was dominant among others in AD of sewage sludge [89]. Metageomic study of microbial variety in sewage sludge of single stage digester by Nelson et al. [107] have revealed that *Chloroflexi* is the abundantly existing microbe. In an analysis of PCR DGGE based microbial community analysis in anaerobic digested cattle manure from single stage digestion, it is revealed that cattle manure was dominated by *Clostridiumglycolicum, Clostridium lituseburense, Clostridium ultunense* etc., [3].In addition, these bacteria, comes under the class, *Gammaproteo* bacteria existed predominantly in single stage digester [108]. These microbial species are reported to be significantly tolerant to toxic materials and extreme environmental conditions [109]. The bacteria that comes under *Bacteroidetes* can secretes hydrolytic enzymes and produce acetic acid during the decomposition of substrates [110]. *Firmicutes* and *Chloroflexi* are the dominant microbial population found in single stage digestion of food waste [107, 111]. The*chloroflexi* was tolerant to high organic loading of substrate [112] and reported *Bacillus infernus* as predominant bacteria in thermophilic 1S-AD treating mixed swine manure and market organic waste [113]. Where it produces formic and lactic acid by fermenting glucose and subsequently use it for its growth. *Porphyromonadaceae,* a doimantant bacterial population in single stage digester was able to produce volatile fatty acids by hydrolysing proteins and carbohydrates [114]. The bacterial abundances are an essential parameter in the formation of methanogenic population in AD. The classes *Clostridia* and *synergistia* were reported to be major classes of bacterial population in 1S-AD of high strength food effluents. Many studies reveled that bacterial populations categorized under the various classes were capable of degrading complex organic matter into simpler compounds and convert lactic acid or acetic acid to hydrogen and carbondioxide [107, 108]. Based on the above, it can be concluded that investigating the changes in microbial communities would provide a significant information on the operation of digester.

8.3.2 Pathways for Single-Stage AD (Microbial Level)

Among the four phases, hydrolytic and acidogenic phases were usually determined as the rate restricting phases in AD process of various wastes, biomass and wastewater. Decomposition of organic matter was performed through different fermentative microbial pathways during these phases. During fermentation, the complex organic matter gets converted into simpler compounds. These compounds in turn utilized as substrates by methanogens [115]. The pathways involved in fermentation includes carbohydrate, protein and aminoacid catabolism, acetic and butyric acid degradation and homoacetogenesis. Similarly, pathway for methanogenesis includes (1) acetoclastic methanogenesis; (2) hydrogenotrophic methanogenesis; (3) methylotrophic methanogenesis; (4) syntrophic hydrogenotrophic methanogenesis. Table 8.2 shows the role of microbes in various metabolic pathways during single stage digestion. The following section describes the various microbial pathways involved in fermentation and methanogenesis. Figure 8.3 displays a schematic representation of pathways for single-stage AD.

Anaerobic fermentative pathways—Carbohydrates, protein and amino acid degradation pathways

In a single stage digestion process, during protein and carbohydrate metabolism, lysis of proteins, and carbohydrates into peptides and sugars were mediated by diverse group of bacteria. The hydrolytic microbes may differ phylogenetically, but comes under the phyla, *Bacteroidetes and Firmicutes* [126]. *Proteobacteria* and *Chloroflexi*

Table 8.2 Role of microbes in various metabolic pathways during single stage digestion

S. No	Digester type	Substrate used	Pathway involved	Role of microbes	References
1	Wet CSTR single stage digester	Swine manure and market biowaste	Aminoacid degradation pathway	*Bacteroidetes*, capable of fermenting proteins and carbohydrates and in particular capable of fermenting aminoacids to acetate	[116]
2	CSTR	Food waste	Stickland reaction	*Proteiniphilum* and *Sedimentibacter* degrade amino acids during anaerobic digestion	[117]
3	CSTR	Food waste	Syntrophic associated amino acid degradation	*VadinBC27* bacterial group degrade amino acid during digestion	[117]
4	CSTR	Food waste	Hydrogenotrophic methanogenesis	*Methanospirillum* and *Methanoculleus* consumes hydrogen in the reactor to form methane	[117]
5	Lab scale continuously stirred biogas reactors	Agricultural residues	Syntrophic acetate oxidation	*Clostridia* are responsible for syntrophic acetate oxidation by coupling with hydrogenotrophic methanogens	[118]
6	CSTR	Sewage sludge	Methylotrophic methanogenesis	The typical genus *Methanomassiliicoccus* and "*Candidatus Methanomethylophilus* degrades the methylcompounds and hydrogen and convert them into methane	[119]
7	CSTR	Sewage sludge	Syntrophic acetogenesis	*Firmicutes*, a sort of syntrophic bacteria that can degrade various substrates and produce VFAs	[119]
8	CSTR	Sewage sludge	Propionate degradation pathway	*Proteobacteria* utilizes propionate and *Chloroflexi* utilizes glucose as substrates	[119]
9	Full scale anaerobic digester	Waste activated sludge	Stickland reaction pathway	*Clostridiales* were the predominant order to degrades amino acids	[120]

(continued)

Table 8.2 (continued)

S. No	Digester type	Substrate used	Pathway involved	Role of microbes	References
10	Full scale anaerobic digester	Waste activated sludge	Hydrolytic and VFA degradation pathway	*Proteobacteria* utilizes glucose and VFAs including propionate, butyrate and acetate	[120]
11	Full scale anaerobic digester	Waste activated sludge	CO_2 reduction pathway	*Clostridium, Treponema, Eubacterium, Thermoanaerobacter* and *Moorella* converts VFA, hydrogen and carbondioxide to acetate	[120]
12	Full scale anaerobic digester	Waste activated sludge	Aceticlastic methanogenesis pathway	*Methanosaeta* utilizes acetate to form methane. *Methanosarcina* utilizes methanol, methylamine and acetate to from methane	[120]
13	Full scale anaerobic digester	Waste activated sludge	Hydrogenotrophic methanogenesis	*Methanospirillum, Methanoculleus* and *Methanoregula* utilizes hydrogen to form methane	[120]
14	Full scale anaerobic digester	Waste activated sludge	Methylotrophic pathway	*Methanococcoides, Methanohalophilus* and *Methanolobus* utilizes methanol (methylated and compounds) and convert them to methane	[120]
15	UASB	Synthetic wastewater containing propionate as sole source carbon	Methylmalonyl coenzyme A pathway Syntrophic β-oxidation	*Pelotomaculum spp.. (P. schinkii,* and *P. propionicicum.)* oxidizes propionate to acetate and butyrate via an integration of two molecules of propionate *Syntrophomonas spp. (S. zehnderi OL-4),* syntrophic fatty-acid-oxidizing bacterium degrades butyrate	[121]
16	Full scale industrial digester	High strength wastewater	Reversible Wood–Ljungdahl pathway (reductive acetyl-CoA pathway)	*Clostridium ultunense* oxidize acetate to provide hydrogen and carbondioxide to syntrophic partner methanogens such as *Methanobacterium*	[122]

(continued)

Table 8.2 (continued)

S. No	Digester type	Substrate used	Pathway involved	Role of microbes	References
17	Single stage thermophilic reactor	Waste activated sludge	Hydrogenotrophic methanogenesis/aceticlastic methanogenesis	*Methanosarcinaceae* cleave acetate and oxidises acetate to H_2	[123]
18	Single stage thermophilic reactor	Waste activated sludge	Syntrophic acetate oxidation	*Thermacetogeniumphaeum* belonging to phylum Firmicutes, *Thermotogalettinga* strain belonging to phylum Thermotogae, These phyla are capable of acetate oxidation	[123]
19	Single stage thermophilic reactor	Waste activated sludge	Hydrogenotrophic methanogenesis	*Methanobrevibacter, Methanothermobacter,* and *Methanobacteriu* consumes hydrogen	[123]
20	Lab scale biogas reactor	Slaughter house waste	Syntrophic acetate oxidation	*Thermacetogeniumphaeum, Clostridium ultunense, SyntrophaceticusschinkiiandTepidanaerobacter Acetatoxydans* degrades acetates	[124]
21	Full scale anaerobic digester	Cattle manure	Aceticlastic methanogenesis pathway	*Methanosarcina* and *Methanoculleus* utilizes acetate to form methane	[125]

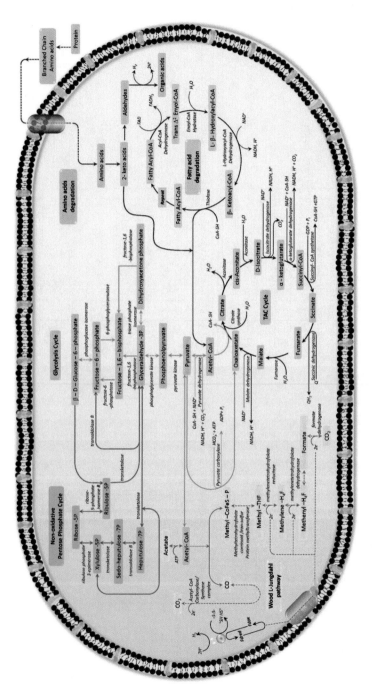

Fig. 8.3 Schematic representation of pathways for single-stage AD

were the dominant microbial population involved in protein and glucose fermentation during single stage digestion in CSTR [127]. While following the hydrolytic and VFA degradation pathway in single stage large scale AD treating sludge the role of *proteabacteria,* which utilizes glucose and VFAs including propionate, butyrate and acetate[128]. A typical glucose fermentative reaction is represented as follows [129]:

$$C_6H_{12}O_6 + 2H_2O \rightarrow 2CH_3COOH + 2CO_2 + 4H_2 \tag{8.3}$$

The acidogenic microbes convert the simple sugars, amino acids, fatty acids into short chain fatty acids, ammonia, hydrogen, alcohols and carbondioxide. The acidogens comes under the phyla *Firmicutes, Bacteroidetes, Proteobacteria, Chloroflexi, and Actinobacteria* [130]. Amino acid metabolism was mediated by acidogenic microbes. The amino acids such as methionine, glycine, cysteine, leucine, isoleucine, threonine, valine, and serine are reported to be the mainly degraded by Stickland reactions [131]. Two distinct pathways were proposed for amino acid fermentation namely Stickland (amino acid pairs were degraded) and syntrophic reaction (degradation of single amino acids through assistance with hydrogen consuming bacterial species). During the degradation of amino acid in single stage CSTR treating food waste. The bacterium, *Proteiniphilum and Sedimentibacter* mediates the degradation of amino acid through Stickland reaction [133]. Contrastingly, *Clostridiales* also act as the predominant organism responsible to degrade amino acids through Stickland reaction [128]. In the process *Bacteroidetes* were able to degrade proteins, carbohydrates and amino acids to acetate through amino acid degradation pathway [8]. Working on single stage AD treating food waste it is documented that syntrophic associated amino acid degradation [93]. They found the bacterium *VadinBC27* in syntrophic association with hydrogen consuming bacteria degraded amino acid.

VFA degradation pathways

VFA, an important intermediate of AD and its subsequent degradation are carried out by vast group of specialized microorganisms. Following are the pathway of those microorganism. It includes Methylmalonyl coenzyme (classical), Smithella (dismutation pathway), CO_2 reduction, syntrophic β oxidation, Reversible Wood–Ljungdahl (reductive acetyl-CoA pathway) and syntrophic acetate oxidation pathway. Among them, Methylmalonyl coenzyme A and Smithella were predominant pathways and were involved in oxidation of propionate. In propionate degradation in single stage UASB by feeding it as a substrate is oxized by *Pelotomaculum spp.. (P. schinkii,* and *P. propionicicum.)* to acetate and butyrate via assimilation [129]. A typical propionate oxidation reaction through methyl melonyl Coenzyme A pathway is represented as follows [129]

$$CH_3CH_2COO^- + 3H_2O \rightarrow CH_3COO^- + HCO_3^- + H^+ + 3H_2 \tag{8.4}$$

In Smithella pathway, *Smithella spp.* utilizes propionate in which propionate was converted to acetate and butyrate through 6C transitional path prior to getting decomposed via β-oxidation [130]. A typical propionate oxidation reaction through

Smithella pathway is represented as follows [130]:

$$CH_3CH_2COO^- + H_2O \rightarrow 3/2CH_3COO^- + 1/2H^+ + H_2 \qquad (8.5)$$

Syntrophic oxidation of butyrate was observed in UASB and by *Syntrophomonas spp. (S. zehnderi OL-4)* [129]. Similarly working on butyrate degradation in CSTR study it is found that degradation happens via β oxidation. A typical butyrate oxidation reaction is represented as follows [130]

$$CH_3CH_2COO^- + 2H_2O \rightarrow 2CH_3COO^- + 2H_2^+ + 2H^+ \qquad (8.6)$$

Propionate degradation pathways in CSTR treating sewage sludge and revealed that the bacterium *Proteobacteria* becomes responsible for it degradation [127]. It is observed further that in single stage AD treating waste activated sludge CO_2 reduction pathway take place [128]. These findings revealed that the bacterium, *Clostridium, Treponema, Eubacterium, Thermoanaerobacter* and *Moorella* converts VFA, hydrogen and carbondioxide to acetate via these pathways. While treating high strength wastewater in single stage full scale industrial reactor, it is Reversible Wood–Ljungdahl pathway for VFA degardation (reductive acetyl-CoA pathway) that occurs [132]. These findings revealed that the bacterium *Clostridium ultunense* oxidize acetate and supplies hydrogen and carbondioxide to its syntrophic partner, *Methanobacterium,* a methanogen. Syntrophic acetate oxidation is another important metabolic pathway responsible for VFA degradation and was evidence in single stage AD treating slaughter house waste [133] and waste activated sludge [134]. Further *Thermacetogenium phaeum* belonging to phylum *Firmicutes* and *Thermotoga lettinga* strain belonging to phylum *Thermotogae,* are capable of acetate oxidation [133]. Whereas the species *Thermacetogenium phaeum, Clostridium ultunense, Syntrophaceticus schinkii* and *Tepidanaerobacter Acetatoxydans* were responsible for the degradation of acetate [134].

Methanogenic pathways

Methanogenic pathways are accountable for the generation of biofuel, biomethane in AD. The pathways involved in methanogenesis are acetoclastic methanogenesis, Syntrophic and hydrogenotrophic methanogenesis. In acetoclastic methanogenic pathway acetate is consumed in a direct way and results in the production of methane and carbondioxide. At first, the substrate acetate is converted to acetyl CoA. The dominant organism responsible for this pathway belongs to the genus, *Methanosarcina* and *Methanosaeta*. The bacterium, *Methanosarcina* consumes less affinity acetate kinase-phosphotransacetylase (PTA) and then converts acetate to acetyl CoA. The bacterium, *Methanosaeta,* consumes greater affinity adenosine monophosphate (AMP)- and forms acetyl CoA synthetase. Finally, the acetyl CoA is transformed to methyl group and then to biomethane. Methane production in HM pathway is achieved via reduction of CO_2 and utilization of H_2 or formate. In case of methylotrophic methanogenesis pathway, methane generation is achieved via degradation of methyl compounds. In hydrogenotrophic methanogenesis syntrophic association between acetate oxidisers and hydrogenotrophic methanogenic microbes convert acetate into methane with the intermediate formation such as hydrogen and carbondioxide. The CO_2 produced during hydrogenotrophic methanogenesis, is further oxidized to intermediate compounds such as formyl, methylene, and methyl groups. The intermediate methyl group is latter transformed to Coenzyme M and generates methyl-CoM. The methyl—CoM is then get converted to biomethane. This process is mediated by methyl coenzyme M reductase (Mcr).

Among these pathways, acetoclastic methanogenesis was considered as the major pathways and was responsible for 70% of methane production [113]. The dominance of hydrogenotrophic/acetoclastic methanogenesic pathway has been observed in single stage thermophilic reactor treating waste activated sludge [133]. Where the bacterium *Methanosarcinaceae* split acetate and oxidizes it to hydrogen. A typical acetoclastic or hydrogenotrophic methanogenic reaction is represented as follows [131]

(i) Acetoclastic methanogenesis

$$CH_3COO^- + H_2O \rightarrow CH_4 + HCO_3^- \qquad (8.7)$$

$$4H_2 + HCO_3^- + H^+ \rightarrow CH_4 + 3H_2O \qquad (8.8)$$

The bacterium *Methanomassiliicoccus* and *Candidatus methanomethylophilus* converts the methyl compounds and hydrogen into methane through methylotrophic methanogenesis pathway [127]. Working on food waste in single stage it is identified that the bacterium *Methanospirillum* and *Methanoculleus* consumes hydrogen in the reactor to form methane during hydrogenotrophic methanogenesis [93]. In case of the bacterium, *Methanosarcina* and *Methanoculleus*it utilizes acetate to form methane during acetoclastic methanogenesis while treating cattle manure in full scale single stage digester [135]. Similarly, it is reported that the bacterium, *Methanosaeta* utilizes acetate to form methane [128]. The bacterium *Methanosarcina* can utilize methanol, methylamine and acetate as substrate to from methane during acetoclastic methanogenesis in a single stage full scale digestion. Hydrogenotrophic methanogenesis and methylotrophic pathways have also been reported in some studies [128]. These findings revealed that the bacterium, *Methanospirillum, Methanoculleus* and *Methanoregula* utilizes hydrogen to form methane during hydrogenotrophic methanogenesis while *Methanococcoides, Methanohalophilus* and *Methanolobus* utilizes methanol (methylated and compounds) to form methane during methylotrophic methanogenesis.

8.3.3 Process Parameters

1S-AD is a sequential technique concerning biodegradation of complex substances and is mediated by several groups of microbial populations. The activity of these microbial population is influenced by several process or operating parameters. Table 8.3 list influence of operational parameters on 1S-AD. These parameters in turn impacts the performance of digestion system and generation of biogas. The fermentative lytic bacteria, acetogens and methanogens could vary extensively in their favorable environment needs. All the reactions (hydrolytic, acidogenic and methanogenic) happen in one reactor is known as single stage digestion. Therefore, a stable process performance and equivalent degradation of substrates must be essential in all the reaction phases. In single stage digestion, typically the needs of methanogenic bacteria must be taken into account with significant consideration and preference. Because these methanogens take lengthy period for regeneration, shows very sluggish growth and are much subtle to ambience conditions when compared to other microbes existing in the mixed populations.

There are some exceptional cases, for example, the hydrolytic phase is the rate restricting phase when treating the substrates rich in cellulose. At the same time, while treating substrates rich in protein, the growth parameters (optimum pH) is found to be equal in all the phases of anaerobic digestion process, in that cases, single stage digestion is found to be satisfactory. The hydrolysis of lipid containing substrate during digestion leads to emulsification which hampers the acetogenesis phase. In such case, thermophilic temperature should be maintained to avoid emulsification. However, the accomplishment of stable performance of single stage digestion process mainly relies upon the rate at which the development and metabolism of all microbial populations takes place. The operational process parameters need to be regulated for enhanced microbial activity, improved digestion efficiency and biogas generation. The influences of such process parameters including OLR, HRT, temperature, mixing, substrates, productivity, and yield on single stage digestion process efficiency are clarified below.

Substrate loading

Organic loading rate (OLR) specifies the concentration of volatile solids (VS) or chemical oxygen demand (COD) to be added into the reactor each day. OLR regulates the microbial activity and conversion efficiency of a digester. HRTis another important parameter which determines the substrate retention time in the reactor. HRTis proportionally inverse to OLR and these two factors are considered essential in determining the stable function of the digester. OLR mainly relies on the type of feedstock fed into the digester [146]. Changes in the concentration of input COD or loading rate could disturb the stability between acidogenic and methanogenic microbes. Many researchers have reported about the influence of OLR and HRTon single stage AD [66] on the basis of pH, alkalinity, solids reduction, biogas production and methanogenic population. To reduce the cost and to augment the potential of 1S-AD, it is feasible to use high OLR. But operation of single stage digestion at high OLR is a great task due to faster acidogenesis and rapid decline of pH. The OLR also depends on HRT. When the OLR is high, the HRT must be satisfactory for the microbial populations to degrade the substrates and increase the methane yield. In addition, at lower HRT, the methanogens get washed away along with the treated effluent [61]. When the OLR was increased above the optimal level, then it could affect the reactor

Table 8.3 Influence of 1S-AD operating parameters and their process efficiency

S. No	Influencing OLR	Factor responsible for inhibition	Type of substrate	Process efficiency	References
1	3 g VS/L d & 7d	Increase in VFA concentration (acidification is reversible at a lower OLR and a higher HRT and the VFA concentration is 4.4 and 3.9 g/L for food waste and municipal solid waste)	Municipal solid waste (OFMSW)	• Methane yield—0.24 m^3/kg VS • Theoretical methane yield—48% • VFA concentration—3.9 g/l • VS reduction—94%	[136]
	4 g VS/L d & 5d		Food processing industry	• Methane yield less than 0.01 m^3/kg VS • Theoretical methane yield—2% • VS reduction—97%	
2	7 g VS/L d	Sudden increase in the VFA concentration decreases the methane yield and VS reduction	Mixed waste (food waste and waste activated sludge in the ration of 1:2)	• TVFA/alkalinity ratio—0.9 • VS reduction decrease to 44% • Methane yield—0.15 L CH$_4$ /g VS$_{added}$	[137]
3	5.5 kg VS/m^3 day	VFA accumulation (pH sharply decrease and VFA concentration increase to 8149 mg/L at mentioned OLR)	Food waste	• Methane yield—421 mL/g VS • Volume of biogas production rate—4.2 L/L d • VS reduction—89%	[138]
4	3.5 g VS/L d & 30 d	VFA accumulation (VFA concentrations was increase to 2302 mg/L)	Mixed food waste (fruit and vegetable waste)	• Biogas production—1.95 L/L d • Methane content—59.7% • Methane yield—385 mL/ g VS	[139]
5	6 g VS/L d	Increase in the concentration acetate under high OLR results in accumulation of VFA and leads to process deterioration	Food waste	• VFA/TA ratio—1.27 • VS reduction decrease to 44% • 50% reduction in methane production,—(0.25 L CH$_4$ /g VS)	[140]

(continued)

Table 8.3 (continued)

S. No	Influencing OLR	Factor responsible for inhibition	Type of substrate	Process efficiency	References
6	10.28 g COD/L d & 11 d	Accumulation of VFA (VFA 7.2 g/L and a drop in pH to 6.85)	Coffee production waste	• VFA/ALK ratio exceeded the limit of 0.3 • TS reduction 29.67% • VS reduction 32.63% • Methane yield—71.63 L CH_4/g TS_0	[141]
7	46 g TVS/l/d & 2d	Changes in pH and VFA can indicate the instability of the reactor (TVFA/alkalinity ratio exceeds the limit of 0.3 and reached 2.73)	Industrial municipal solid waste	• COD removal—9% • TVS removal—29% • Sulphate removal—71% • TVFA/alkalinity ratio—2.73 • Methane content—50% • Specific methane production—0.06 m^3 CH_4/kg TVS	[142]
8	15 kg TVS/m^3 & 2d	Accumulation of VFA in this reactor	Waste activated sludge	• TVS removal—34% • Methane content—62% • Biogas production—0.07 m^3/d • Biogas production rate—0.05 m^3/m^3 d	[143]
9	3.5 kg TVS/m^3 d & 2d	Accumulation of VFAs, and in particular, propionic acid, possibly indicating a potential change in the metabolic pathways	Mixed food waste	• Specific Gas Production—0.75 m^3 biogas/kg VS • Specific Methane Production—0.45 m^3 CH_4/kg VS • Gas Production Rate—2.5 m^3 biogas/($m^3_{reactor}$ d)	[144]
10	3.5 g COD/L d	VFA accumulation results in pH drop and ultimately cause process failure	Municipal sludge	• VS reduction—26.3% • Total COD reduction—30.8% • Methane yield—0.16 L CH_4/g VS • Methane production—15.6 L/d	[145]

performance by decreasing the methane yield, COD and VS removal efficiency. In addition, a greater amount of suspended solids could eluted along with the treated effluent (biomass wash out) which causes imbalances in the system. It is found that the increase in OLR significantly increases the concentration of total solids in the reactor which may reduce the mass transfer efficiency and decrease the biogas yields [61]. These findings revealed that methane production percentage decreased by 43% on increment of substrate loading from 1 to 6 g VS/L/d. The OLR remains inadequate at 1–4 kg VS/m³/d for single stage digestion [66]. At higher OLR process inhibition was caused by accumulation compounds such as VFA, ammonia (inhibitory) and phenol (toxic) in the system. VFA accumulation irreversibly acidify the single stage reactor and it leads to reactor souring. At high OLR, the VS removal efficiency and methane yield decreases and this can be owing to the existence of non-biodegradable organics in the digested residue, which remains as such without getting converted into VFA [66]. While working on single stage mesophilic AD for biodegrading food waste at high OLR, the abundance of acidogens and syntrophic VFA oxidizers over hydrogenotrophic methanogens was found [92]. At low HRT, the ratio of Eubacteria to Archaea was increased. This indicates more wash out of methanogen proportions over eubacteria. As a result, increase in acetogenic population occur which increase the VFA production and decreases the predominance of methanogens. Further an increment in substrate loading from 30.7 g to 46.0 g TVS/L/d adversely affect the performance of AD [147]. For instance, souring of a single stage digestion treating fruits and vegetable wastes, while increasing substrate loading from 3 to 3.5 g VS/L/d was observed in a study [67].

Similarly, VFA build up and acidification of digester at high operating OLR and low HRT is observed [60]. It is also noticed a decline in pH of the reactor because of VFA build up [60]. This leads to restriction of methanogens and pH drop make them unable to transform VFA into methane. The pH of single stage digester should be maintained between 6.8–8.0 (optimal range for methanogens) in order to reduce the inhibition effect [148]. A drop in pH (from 7.2–5.8) was found while treating fruits and vegetable waste at higher OLR of 3.5 g VS/L/d [67]. It is found that a drop in pH while increasing the substrate loading from 3 to 12 kg VS/m³/d. Many studies have reported that increasing the OLR beyond reactor threshold level subsequently decrease the biogas production in a single stage digestion [92].

Similar decrease in methane production was obtained at higher OLR. A decrease in specific biogas production and methane production was observed while increasing the substrate loading from 8–12 kg VS/m³/d [92]. These findings revealed that over-load of the digester causes accumulation of VFA and inhibition of acetoclastic and hydrogenotrophic methanogens and this could be the reason for decreased biogas generation.

Success of anaerobic digestion depends on obtaining maximum operational threshold organics loading. Researchers have optimized a ultimate OLR of 30.7 g VS/L/d in a single phase dry digester by stepwise increment of substrate load from 5.7 g VS/L/d [147]. As a result, the rate of methane generation was increased from 1.5 to 6.9 L CH_4/L/d. It is also reported further that a destabilization of the digestion system and wash out of methanogenic population were happened while increasing the substrate loading beyond the threshold level to 46 g VS/L/d. At greater substrate load (46 g VS/L/d), it has been observed that a sharp fall in COD and VS removal efficiency of the reactor occur. In addition to influence on pH, VS removal and biogas yield, OLR also showed significant influence on microbial communities in single stage digestion. Thus, OLR is an important operating tool for single stage digestion and the optimum OLR may vary depending upon the digester type, substrate used, nutritional requirements, toxic compounds etc.

Temperature

Temperature is an essential environmental factor that impacts the existence of microbial population and mediates the digestion performance. Based on temperature AD was classified into three types namely psychrophilic (10–20 °C), mesophilic (25–37 °C) and thermophilic (55–65 °C). Psychrophilic AD was popular during 1980s. Later, psychrophilic based AD lost its popularity due to its sow metabolic activity. Mesophilic based AD is found to be prevalent in recent times on the other hand, thermophilic based AD also getting attention of process engineers as it offers many benefits. For example, thermophilic single stage digesters provide greater yield, enhanced biodegradation, improved biogas productivity, elimination of pathogens reduced contamination and lesser retention period [149]. The major drawback of single stage thermophilic digesters are its thermal based negative influence on the growth of hydrogenotrophic methanogenic population and higher energy requirement [150]. Mesophilic single stage digesters are operated with healthier microbial population and can withstand drastic changes in environmental conditions. In addition, these digesters are easy to handle provide stable performance, and demand low capital investment. Main demerits are elevated retention time and reduced gas yield. On the basis of composition of substrate and digester types, optimal process temperature differs but it must be sustained constantly for increased methane production [60]. Numerous authors have investigated about the impact of both mesophilic and thermophilic temperature on single stage digestion [151–153]. It has been reported that increasing the temperature during AD would promote the process rate and achieve higher organic loading rate at the same time it will not affect the substrate reduction efficiency [154]. For example, Mesophilic and thermophilic single stage digester to treat palm oil wastewater [155]. In that study it is reported that comparatively a higher methane production rate in thermophilic (4.66 L/L/d) than mesophilic digestion (3.73 L/L/d) have occurred. In contrast to this, another study observed that mesophilic single stage digester surpassed the performance of thermophilic single stage digestion and found a higher specific methane production of 188.3 ± 31.62 mL/g VS in mesophilic single stage digester when compared to thermophilic single stage digester (90.26 ± 17.09 mL/g VS) [155]. In addition, the mesophilic condition also enhanced the organic removal efficiency. A higher VS removal of 83% was achieved in mesophilic digester when compared to thermophilic digester (65%). This could be due to thermal effect at elevated temperature inhibition of acetate consumers (acetoclastic methanogenesis or syntrophic acetate oxidizing microorganisms) occur and this leads to VFA build up in the reactor. Further, a study supported the above outcome in which the authors have observed enhanced action of hydrogenotrophic methanogens at mesophilic condition when compared to thermophilic digester [156]. Based on the above reports it can be concluded that higher temperature upto certain range could improve the rate of hydrolysis and digestion process. In addition, thermophilic microbes are more subtle to slight temperature alterations and therefore many single stage reactors are reported to be operated under mesophilic conditions [157].

Substrates, productivity and yield

Substrate is an essential parameter and it determine the efficiency of AD [68]. Substrate originating from various sources could differ widely and considerably in its characteristics and composition. The chosen substrate must satisfy the critical nutrients requirement of the microbial population in the digester. It should provide energy and components essential for biosynthesis of new microbes [63]. The substrate must also possess vitamins and other vital nutrients essential for action of microbial enzymes. The composition of substrate imparts a significant task in biodegradation and impacts the value of digested residue. On the basis of substrate composition, it has been reported that long chained carbon containing compounds (cellulose) takes time to degrade [158]. If the substrates rich in glucose are used for digestion, then the problem of VFA accumulation will be considerably lower. In case of AD of agricultural residues, lignin a major constituent of agricultural residues remains as recalcitrant and hinder the AD efficiency of digestion.

The nature of substrates directly influences methane productivity and organic removal efficiency of 1S-AD. In addition, the substrates methane yielding efficiency also depends on process temperature [88]. While investigating methane production potential of cow, pig manure and chicken wastes at varying temperature have found that at mesophilic temperature cow dung showed higher methane potential than chicken waste [159]. Various substances which includes municipal waste, food waste, agro waste, aquatic biomass, industrial wastes and wastewater can be used as substrates in single stage digestion [15, 68]. Among the substrate food waste is a potential substrate for AD and has been heavily exploited for biogas production [59]. Food wastes have been used as substrate in thermophilic 1S-AD at 28 days HRT and achieved biogas with 74% methane [160]. Similarly, working on thermophilic 1S-AD of food processing waste in CSTR have been found [60], 94% organics removal and 84% methane production. As food waste contain heavier fraction of biodegradable organics, it can be easily transformed into VFA in short period of time. This leads to VFA build up and decline in pH, which subsequently affect methanogenesis [159]. Accordingly, codigesting food waste with other substrate was proved to be efficient. Codigestion of food waste with piggery wastewater in single stage CSTR, leads to a greater biomethane production of 396 mL/g VS[136].

Agricultural residues are reported to be an appropriate substrate for bioenergy generation owing to their adequate availability, greater methane yield potential and cheaper prices. These include corn stover, rice straw, wheat straw and crop wastes etc. When corn stock is used as a substrate in single stage UASS reactor at a HRT, OLR and temperature of 18 days, 2.5 g VS/L/d and 39 °C respectively, as operational condition they have achieved methane yield and content of 184 mL/gVS and 54.4% was achieved [75]. The crop wastes have predominant amount of carbohydrates such as cellulose, hemicellulose and intact compounds like lignin which resist to undergo biodegradation [159]. In those cases, pretreatment is required to disintegrate the intact structure to enhance biogas production. For example, in a study rice straw is pretreated with sodium hydroxide in 1S-AD and achieved enhanced biogas production [137]. Municipal waste is rich in hydrolysable organics and considered to be the potential substrate for AD. For instance, mesophilic 1S-AD of municipal waste activated sludge in a CSTR working with an operational condition of 15 days retention time and 15 g COD/L/d OLR leads to biomethane production of 455 mL/g VS [66]. In addition to the above, industrial waste, industrial wastewater and aquatic biomass can also be utilized as substrates in 1S-AD.

A coffee processing wastewater in single stage EGSBR at temperature, retention time and OLR of 25 °C, 6 days and 3 gCOD/L/d provided a biomethane production of 300 mL/g COD [161]. Lipid is used to extract microalgal biomass as substrate in 1S-AD under mesophilic condition at 15 days HRT and 2 gVS/L/d OLR and achieved a biomethane production of 195 mL/g VS [103].

Mixing

In AD, mixing holds the solid contents in a suspended form and maintains the substrate of a digester in a homogenized condition along with microbial population [116, 162]. In addition, mixing is also used to sustain temperature and to avoid development of crust coating and solids deposit in a digester. In single stage digestion, the choice of mixing plays an important role in efficient substrate mixing and power utilization. Commonly three kinds of mixing methods are employed in 1S-AD which includes mechanical mixing, pneumatic mixing, and hydraulic mixing [138, 163]. Researchers have different view on mixing methods and its impact on methane production. Mechanical mixing is done through baffles and are made functional either by manual nor electrical mode. In Europe almost fifty percent of the single stage full scale CSTR digesters are operated with mechanical mixers [139]. Mechanical mixing not only enhance the interaction of microbes and substrates but also minimizes the stratification in the single stage digestion. Investigating the effect of mechanical mixing in single stage CSTR treating OFMSW at mesophilic temperature, mechanical mixing showed a greater methane yield of 1.324 L/L/d [164]. Similarly, working in a 1S-AD biodegrading food waste under mesophilic condition. Mechanical mixing enhanced methane production [165]. Gas recirculation is another method of mixing, where anaerobically produced biogas was compressed and used to mix reactor contents. It improves substrate microbe interaction [197. The impact of gaseous recirculation on the stability of 1S-AD operated at 16.2 day HRT and 3.1 g TS/L/d OLR, ascertains that gas mixing enhanced biogas production when compared to unmixing [18]. Recirculation of liquid/leachate is another method of mixing which enhances the substrate microbial interaction. The recirculation method of mixing minimizes the extra energy spent towards mechanical mixers. When leachate circulation is used as mixing in single stage lab scale reactor treating domestic waste under mesophilic condition, a higher methane percentage of 60% is obtained [19]. Similarly use of sludge recirculation as mixing in a prototype mesophilic 1S-AD treating food waste yielded higher methane percentage of 61.6% [19]. The parameters such as shear intensity, power consumption, mode of mixing and mixing time influences the single stage anaerobic digestion efficiency, and the microbial populations in 1S-AD. Mixing intensity is an important parameter that influence the activity of bacterial populations in a single stage digester and biodegradation potential. For instance, greater mixing intensity could affect the formation of flocs and the syntrophic association among hydrogenotrophic and acetogenic microbial population. This results in inhibition of AD by partial pressure of hydrogen and accumulation of volatile fatty acids [20]. While studying the impact of mixing intensity in prototype mesophilic single stage digester treating food waste with 8% total solids (TS), methane yield of 0.83 $m^3/m^3/d$ at a mixing intensity of 0.033 L/min L (30 min twice a day) has been attained [21]. On a study gas mixing was employed with an injection velocity of 7.6 m/s to treat dairy manure with 5.4% TS in a mesophilic single stage digester. During the treatment the investigation it is noticed that floc destruction due to rupture of bubbles and shear force created in the inlet. It was found that mode and duration of mixing can impact the stability of 1S-AD efficiency [21]. The mode of mixing is of three types: intermittent mode, continuous mode, and non-mixed conditions. The modes of mixing in mesophilic single stage pilot scale CSTRs treating dairy manure with 6.1% TS. This study revealed that intermittent mode of mixing minimizes the energy requirement and cost. In addition it also enhances the production of methane in comparison with other mode of mixing [166]. The influence of mixing time 10, 5 and 2 min in a single stage full scale CSTR have been evaluated. The outcome implied that by decreasing the mixing time from 10 to 5 min minimizes 53% of energy utilization and 2 min minimizes the 81% of energy utilization [166].

8.3.4 Mass and Heat Transfer

Mass and heat transfer are the crucial parameters that determines the digestion efficiency and stability of 1S-AD. The mass transfer is defined as stwo types: convection and diffusion. Similarly, heat transfer explains about the transport of heat and temperature gradients within the digester. The heat transfer is of three types: convection, conduction and radiation. Mixing, substrate solids concentrations, substrate diffusion within the biomass are the important parameters that affect the mass transfer efficiency in single stage digestion [23, 24, 167]. It has been reported that mixing enhance the mass transfer efficiency in single stage digestion [23, 25, 167]. The release of gas from digested liquor in single stage digesters working under intermittent mode of mixing could be increased upto 70% [25, 138]. The release of gas from digested liquor was hampered at unmixed conditions but at the same time mass transfer of gas from liquid to gas phase has been enhanced through mixing. Similarly, it is reported that mechanical mixing enhanced the homogenization of substrate and achieved better mass transfer at elevated organic loading [23]. Upon investigating the effect of mixing on mass transfer, it is found that mechanical stirring enhanced gas releasing capacity and mass transfer [168]. Solids concentration of substrate also influences the mass transfer efficiency in single stage digester [24]. Mixing is essential for increased methane production in single stage digester having high TS concentration. In case of non-mixing circumstances, mass transfer is probably mediated through diffusion (related to medium permeability and amount of water). Mass transfer coefficient is an important mathematical parameter and can be calculated as follows:

$$K_{La} = \frac{Q_v}{r([gas]_L/[gas]_{L^*} - 1)} \tag{8.9}$$

where, k_La is the overall mass transfer coefficient (h^{-1}); K_H is the Henry's law constant (moles/L); Q_v is the volumetric gas production rate ((Q_g/V_L)L of gas per L of reactor volume); R is the ideal gas constant (8.314 Pa mole^{-1} K $-$ 1); T is the temperature (Kelvin); V_L is the liquid phase volume; V_g is the gas phase volume; $[gas]_L$, concentration of solubilized gaseous compounds in the digester (generally equivalent to the effluent concentration) (moles/L); $[gas]_{L^*}$, concentration of solubilized gaseous compounds at thermodynamic equilibrium (moles/L). Higher TS concentration and lower mass transfer coefficient reportedly decrease cumulative methane production in single stage batch digester.

The limitation in mass transfer could be due to presence of inorganic carbon (carbon dioxide), soluble methane and soluble hydrogen. Increased inorganic carbon limits the release of soluble compounds and increases the organic acids quantity and resulted in inhibition of cumulative methane production. For example, in a study have reported a decrement in mass transfer efficiency and methane production in a single stage wet digester treating high solids is reported [66]. In case of upflow anaerobic reactor, mass transfer in anaerobic granules is essential for enhanced production of biomethane. In granular based single stage reactors, the biodegradability potential of digesters depends on substrate mass transfer capability into the granules [167]. Internal diffusion of substrate within the granules is rate restricting phase in methanogenesis than external diffusion (passage of substrate from external medium into the granules. The molecular diffusion within the granule was calculated as follows [167]

$$F_{MD} = 4\pi \, r^2 D_M C \, \varepsilon/\phi \tag{8.10}$$

where r is the granule radius (cm), D_M is the substrates diffusivity in water (cm^2/s), C is the substrate concentration (mg/L), and ε is porosity (dimensionless). The convective diffusion rate inside the granule was calculated as follows [167]

$$FCD = 2\pi r^2 \eta CU \tag{8.11}$$

where, η is the granule fluid collection efficiency (dimensionless) and U is superficial liquid velocity (cm/s). Researchers have contrasting results and different views on the mass transfer efficiency in biofilms and granules for methane generation. The mass transfer of substrate into biomass can be enhanced greatly if surface to volume ratio of biomass is increased [169]. Researchers have reported that single stage fixed film anaerobic reactor produce very tiny biological particles with optimum surface to volume ratio. Mainly fixed film bioreactors are found to be liable to surface area limitations. A higher mass transfer was observed while operating fluidized bed reactor having bricks as carrier. Carrier material facilitate biomass to achieve higher surface to volume ratio [170]. Some researchers have reported that granules were impermeable and mass transfer could be mediated by molecular diffusion and not in convective mode of transfer. Some researchers have inferred that mode of mass transfer was convective in granules [171, 172]. It is observed that permeability in granules through convection increased the biomethane generation in single stage upflow reactors. Further the convection would increase the mass transfer rate inside different granular sizes ranges from small (0.5–1 mm), medium (1.5–2 mm) to large (3–3.5 mm) [172]. The mass transfer through convection of granular sludge were calculated to be 3 m/h and it significantly enhances the production of biogas upto 4, 1.5 and 1.3 folds in small, medium and larger granules. In addition to mass transfer, heat transfer is an essential factor in single stage digester. Anaerobic microbial consortia are reported to be sensitive to the digestion temperature [173]. Maintaining constant optimum temperature within the digester enhance biogas production.

Factors that play an important role in heat transfer and are (i) weather condition, (ii) solar irradiation, (iii) location of digester, (iv) digester design, (v) influent temperature and its (vi) flow rate. There are three different modes of heat transfer that occur in single stage complete mix digester [174]. They are convective heat transfer that happen between mixing liquid and the digester walls, conductive heat transfer that happen between inner and outer region of the digester and solar irradiation. In addition, mechanical mixing also has greater impact on heat transfer efficiency [174]. During mechanical mixing, the speed of impeller or agitator improves convective mode of heat transfer within the digester. The mixing has meager influence on heat loss from the reactor as conduction heat transfer is relied on thickness of reactor wall, conductivity of wall and gradient of temperature. Also it is observed that in single stage plug flow reactors, the heat gained is found to be higher than heat loss across the latitude of digester. The heat gained through solar energy is enough to compensate the heat loss and it prevent the additional energy input. In summer time, the methane production increased with increase in heat fluxes but at winter times, the methane production decreased due to lesser heat fluxes. The temperature of digester can be regulated by adjusting the temperature of influent. A nonlinear increase in methane yield could be obtained at increased digester temperature. From the above, it can be concluded that there is an inherent link exist between methane yield and mixing, mass and heat transfer.

8.3.5 Advantages and Limitations

Single stage anaerobic digesters are considered to be attractive due to its easiness in configuration and working [175]. About 95% of the installed pilot scale anaerobic digesters are of single stage system [26]. Technical failure is very low in 1S-AD. The cost of 1S-ADs are very less when compared to other form of digesters [173]. In addition, the substrate degradation is very fast due to recirculation of digested residue. In pilot scale single stage system, inexpensive instruments are employed to handle and transport slurry. Among the single stage digesters, the single stage batch reactors are comparatively simple in operation and robust in nature. A major advantage of single stage digester is its elevated reaction rate. Since, reaction rate of single stage batch reactors are high, it yield very high methane [27]. The single stage wet digesters are highly tolerant to unwanted materials such as stone pieces, wood chips and glass particles. Moreover, the generation of inhibitors during digestion is the major disadvantage associated with single stage digestion. It can be overcome by diluting feed material with fresh water [27]. The single stage dry digesters are more robust in nature, and they possess greater retention of biomass. The pretreatments employed for dry digester are found to be cheaper. An important benefit of thermophilic dry single stage digestion system is hygienization of substrates. This results in the production of well digested and pathogen eliminated digestate.

The single stage digesters possess some limitations. In single stage, the acidogenic and methanogenic microbial population have to sustain under similar environmental conditions, even though they differ in their growth parameters, nutrient requirements, and ecological conditions such as temperature, pH etc. [143]. For example, though the optimum pH of 1S-AD ranges from 6.8–7.4, the optimal pH of acetogenesis (pH 5.3–5.7) and methanogenesis (pH 6.5–7.5) differ significantly. This leads to decrease in the productivity of digester [176]. The aim of pretreatments is to eliminate inhibitory and other unwanted compounds during 1S-AD. It may leads to loss of organics (15–25% of volatile solids) which in turn causes decrement in methane production [177]. Even though the pretreatments methods in single stage digestion system make the digestibility of substrates easier, they inexorably causes increment in operational and capital costs. In some cases, pretreatment results in the formation of toxic products [178]. In AD, nearly 2/3rd of the biogas is generated from acetic acid through methanogenesis. Among the total methanogenic population syntrophic acetoclastic methanogens are found to be slow growers and its methane production potential in a single stage mesophilic digester is often disturbed due to the presence of excess amount of propionate and butyrate [179].

The major technical issue associated with single stage digestion is the deposition of thicker particles in the base of reactor and the development of free scum coating at the uppermost of digester. The thickened fractions at the bottom could impair the baffles and top scum layer influence mixing. Therefore, it is essential to remove the settled thicker portion and layer of scum periodically, which otherwise cause reduction methane productivity [176]. The digested slurry comes out from single stage anaerobic digester is not completely devoid of moistness and it require drying before it can be utilized as fertilizer. In such cases, it demand an ample amount of energy for drying [176]. Other minor issues include requirement of larger quantity of water for feed dilution and foam formation [173].

References

1. Van Hanndel C, Lettinga G (1994) Anaerobic sewage treatment: A practical guide for regions with a hot climate. Wiley publishers, U.K.
2. Tomei MC, Braguglia CM, Mininni G (2008) Anaerobic degradation kinetics of particulate organic matter in untreated and sonicated sewage sludge: role of the inoculum. Bioresour Technol 99:6119–6126
3. Xiao B, Qin Y, Zhang W, Wu J, Qiang H, Liu J, You Li Y (2018) Temperature-phased anaerobic digestion of food waste: A comparison with single-stage digestions based on performance and energy balance. Bioresour Technol 249:826–834
4. Voelklein MA, Jacob A, O'Shea R, Murphy JD (2016) Assessment of increasing loading rate on two-stage digestion of food waste. Bioresour Technol 202:172–180
5. De Gioannis G, Muntoni A, Polettini A, Pomi R, Spiga D (2017) Energy recovery from one-and two-stage anaerobic digestion of food waste. Waste Manag 68:595–602
6. Lee D-Y, Ebie Y, Xu K-Q, Li Y-Y, Inamori Y (2010) Continuous H2 and CH4 production from high-solid food waste in the two-stage thermophilic fermentation process with the recirculation of digester sludge. Bioresour Technol 101:S42–S47
7. Leite WRM, Gottardo M, Pavan P, Belli Filho P, Bolzonella D (2016) Performance and energy aspects of single and two phase thermophilic anaerobic digestion of waste activated sludge. Renew Energy 86:1324–1331
8. Schievano A, Tenca A, Scaglia B, Merlino G, Rizzi A, Daffonchio D, Oberti R, Adani F (2012) Two-stage vs single-stage thermophilic anaerobic digestion: comparison of energy production and biodegradation efficiencies. Environ Sci Technol 46:8502–8510
9. Park Y, Hong F, Cheon J, Hidaka T, Tsuno H (2008) Comparison of thermophilic anaerobic digestion characteristics between single-phase and two-phase systems for kitchen garbage treatment. J Biosci Bioeng 105:48–54
10. Chu C-F, Li Y-Y, Xu K-Q, Ebie Y, Inamori Y, Kong H-N (2008) A pH-and temperature-phased two-stage process for hydrogen and methane production from food waste. Int J Hydrogen Energy 33:4739–4746
11. Thompson RS (2008) Hydrogen production by anaerobic fermentation using agricultural and food processing wastes utilizing a two-stage digestion system
12. Dong L, Zhenhong Y, Yongming S, Longlong M (2011) Anaerobic fermentative co-production of hydrogen and methane from an organic fraction of municipal solid waste. Energy Sour, Part A Recover Util Environ Eff 33:575–585
13. Gómez X, Cuetos MJ, Prieto JI, Morán A (2009) Bio-hydrogen production from waste fermentation: mixing and static conditions. Renew Energy 34:970–975
14. Ren Y, Yu M, Wu C, Wang Q, Gao M, Huang Q, Liu Y (2017) A comprehensive review on food waste anaerobic digestion: Research updates and tendencies. Bioresour Technol 247:1069–1076
15. Kothari R, Pandey AK, Kumar S, Tyagi VV, Tyagi SK (2014) Different aspects of dry anaerobic digestion for bio-energy: An overview. Renew Sustain Energy Rev 39:174–195
16. De Mes TZD, Stams AJM, Reith JH, Zeeman G (2003) Methane production by anaerobic digestion of wastewater and solid wastes. Bio-Methane & Bio-Hydrogen 2003:58–102
17. Li W-W, Yu H-Q (2011) From wastewater to bioenergy and biochemicals via two-stage bioconversion processes: a future paradigm. Biotechnol Adv 29:972–982
18. Karim K, Thoma GJ, Al-Dahhan MH (2007) Gas-lift digester configuration effects on mixing effectiveness. Water Res 41:3051–3060
19. Sponza DT, Ağdağ ON (2004) Impact of leachate recirculation and recirculation volume on stabilization of municipal solid wastes in simulated anaerobic bioreactors. Process Biochem 39:2157–2165
20. Ratanatamskul C, Saleart T (2016) Effects of sludge recirculation rate and mixing time on performance of a prototype single-stage anaerobic digester for conversion of food wastes to biogas and energy recovery. Environ Sci Pollut Res 23:7092–7098

21. Wu B (2014) CFD simulation of gas mixing in anaerobic digesters. Comput Electron Agric 109:278–286
22. Rico C, Rico JL, Muñoz N, Gómez B, Tejero I (2011) Effect of mixing on biogas production during mesophilic anaerobic digestion of screened dairy manure in a pilot plant. Eng Life Sci 11:476–481
23. Wu B (2010) CFD simulation of gas and non-Newtonian fluid two-phase flow in anaerobic digesters. Water Res 4:3861–3874
24. Abbassi-Guendouz A, Brockmann D, Trably E, Dumas C, Delgenès J-P, Steyer J-P, Escudié R (2012) Total solids content drives high solid anaerobic digestion via mass transfer limitation. Bioresour Technol 111:55–61
25. Ong HK, Greenfield PF, Pullammanappallil PC (2002) Effect of mixing on biomethanation of cattle-manure slurry. Environ Technol 23:1081–1090
26. Mata-Alvarez J (2002) Biomethanization of the organic fraction of municipal solid wastes. IWA publishing
27. Gonzalez-Martinez A, Garcia-Ruiz MJ, Rodriguez-Sanchez A, Osorio F, Gonzalez-Lopez J (2016) Archaeal and bacterial community dynamics and bioprocess performance of a bench-scale two-stage anaerobic digester. Appl Microbiol Biotechnol 100:6013–6033
28. Kavitha S, Jayashree C, Adish Kumar S, Yeom IT, Banu JR (2014) The enhancement of anaerobic biodegradability of waste activated sludge by surfactant mediated biological pretreatment. Bioresour Technol 168:159–166
29. Banu JR, Kannah RY, Kavitha S, Gunasekaran M, Kumar G (2018) Novel insights into scalability of biosurfactant combined microwave disintegration of sludge at alkali pH for achieving profitable bioenergy recovery and net profit. Bioresour Technol 267:281–290
30. Banu JR, Eswari AP, Kavitha S, Kannah RY, Kumar G, Jamal MT, Saratale GD, Nguyen DD, Lee D-G, Chang SW (2019) Energetically efficient microwave disintegration of waste activated sludge for biofuel production by zeolite: Quantification of energy and biodegradability modelling. Int J Hydrogen Energy 44:2274–2288
31. Kannah RY, Kavitha S, Banu JR, Karthikeyan OP, Sivashanmugham P (2017) Dispersion induced ozone pretreatment of waste activated biosolids: Arriving biomethanation modelling parameters, energetic and cost assessment. Bioresour Technol 244:679–687
32. Akbulut A (2012) Techno-economic analysis of electricity and heat generation from farm-scale biogas plant: Çiçekdağı case study. Energy 44:381–390
33. Micolucci F, Gottardo M, Pavan P, Cavinato C, Bolzonella D (2018) Pilot scale comparison of single and double-stage thermophilic anaerobic digestion of food waste. J Clean Prod 171:1376–1385
34. Massanet-Nicolau J, Dinsdale R, Guwy A, Shipley G (2015) Utilising biohydrogen to increase methane production, energy yields and process efficiency via two stage anaerobic digestion of grass. Bioresour Technol 189:379–383
35. Stabnikova O, Liu XY, Wang JY (2008) Digestion of frozen/thawed food waste in the hybrid anaerobic solid–liquid system. Waste Manag 28:1654–1659
36. Dareioti MA, Kornaros M (2015) Anaerobic mesophilic co-digestion of ensiled sorghum, cheese whey and liquid cow manure in a two-stage CSTR system: effect of hydraulic retention time. Bioresour Technol 175:553–562
37. Kongjan P, O-Thong S, Angelidaki I, (2011) Performance and microbial community analysis of two-stage process with extreme thermophilic hydrogen and thermophilic methane production from hydrolysate in UASB reactors. Bioresour Technol 102:4028–4035
38. Maspolim Y, Zhou Y, Guo C, Xiao K, Ng WJ (2015) Comparison of single-stage and two-phase anaerobic sludge digestion systems–performance and microbial community dynamics. Chemosphere 140:54–62
39. Shin SG, Han G, Lim J, Lee C, Hwang S (2010) A comprehensive microbial insight into two-stage anaerobic digestion of food waste-recycling wastewater. Water Res 44:4838–4849
40. Trisakti B, Irvan M, Turmuzi M (2017) Effect of temperature on methanogenesis stage of two-stage anaerobic digestion of palm oil mill effluent (POME) into biogas. In: IOP Conference series materials science and engineering. IOP Publishing 206:12027

41. Wan S, Sun L, Sun J, Luo W (2013) Biogas production and microbial community change during the Co-digestion of food waste with chinese silver grass in a single-stage anaerobic reactor. Biotechnol Bioprocess Eng 18:1022–1030

42. Song Y-C, Kwon S-J, Woo J-H (2004) Mesophilic and thermophilic temperature co-phase anaerobic digestion compared with single-stage mesophilic-and thermophilic digestion of sewage sludge. Water Res 38:1653–1662

43. Chamy R, Vivanco E, Ramos C (2011) Anaerobic mono-digestion of Turkey manure: efficient revaluation to obtain methane and soil conditioner. J Water Resour Prot 3:584

44. Liu X, Wang W, Shi Y, Zheng L, Gao X, Qiao W, Zhou Y (2012) Pilot-scale anaerobic co-digestion of municipal biomass waste and waste activated sludge in China: effect of organic loading rate. Waste Manag 32:2056–2060

45. Moon HC, Song IS (2011) Enzymatic hydrolysis of foodwaste and methane production using UASB bioreactor. Int J Green Energy 8:361–371

46. Agyeman FO, Tao W (2014) Anaerobic co-digestion of food waste and dairy manure: effects of food waste particle size and organic loading rate. J Environ Manage 133:268–274

47. Zhang C, Xiao G, Peng L, Su H, Tan T (2013) The anaerobic co-digestion of food waste and cattle manure. Bioresour Technol 129:170–176

48. Aslam M, McCarty PL, Shin C, Bae J, Kim J (2017) Low energy single-staged anaerobic fluidized bed ceramic membrane bioreactor (AFCMBR) for wastewater treatment. Bioresour Technol 240:33–41

49. Youngsukkasem S, Akinbomi J, Rakshit SK, Taherzadeh MJ (2013) Biogas production by encased bacteria in synthetic membranes: protective effects in toxic media and high loading rates. Environ Technol 34:2077–2084

50. Li Y, Gao M, Hua D, Zhang J, Zhao Y, Mu H, Xu H, Liang X, Jin F, Zhang X (2015) One-stage and two-stage anaerobic digestion of lipid-extracted algae. Ann Microbiol 65:1465–1471

51. Moset V, Ottosen LDM, Xavier C de AN, Møller HB. Anaerobic digestion of sulfate-acidified cattle slurry: one-stage versus two-stage. J Environ Manage 2016;173:127–33.

52. Puebla YG, Pérez SR, Hernández JJ, Renedo VS-G (2013) Performance of a UASB reactor treating coffee wet wastewater. Rev Ciencias Técnicas Agropecu 22:35–41

53. Yousefzadeh S, Ahmadi E, Gholami M, Ghaffari HR, Azari A, Ansari M, Miri M, Sharafi K, Rezaei S (2017) A comparative study of anaerobic fixed film baffled reactor and up-flow anaerobic fixed film fixed bed reactor for biological removal of diethyl phthalate from wastewater: a performance, kinetic, biogas, and metabolic pathway study. Biotechnol Biofuels 10:139

54. Cruz-Salomón A, Ríos-Valdovinos E, Pola-Albores F, Meza-Gordillo R, Lagunas-Rivera S, Ruíz-Valdiviezo VM (2017) Anaerobic treatment of agro-industrial wastewaters for COD removal in expanded granular sludge bed bioreactor. Biofuel Res J 4:715–720

55. Li C, Tabassum S, Zhang Z (2014) An advanced anaerobic expanded granular sludge bed (AnaEG) for the treatment of coal gasification wastewater. RSC Adv 4:57580–57586

56. Nkemka VN, Murto M (2013) Biogas production from wheat straw in batch and UASB reactors: the roles of pretreatment and seaweed hydrolysate as a co-substrate. Bioresour Technol 128:164–172

57. España-Gamboa EI, Mijangos-Cortés JO, Hernández-Zárate G, Maldonado JAD, Alzate-Gaviria LM (2012) Methane production by treating vinasses from hydrous ethanol using a modified UASB reactor. Biotechnol Biofuels 5:82

58. Varol A, Ugurlu A (2017) Comparative evaluation of biogas production from dairy manure and co-digestion with maize silage by CSTR and new anaerobic hybrid reactor. Eng Life Sci 17:402–412

59. Zhang Y, Banks CJ, Heaven S (2012) Co-digestion of source segregated domestic food waste to improve process stability. Bioresour Technol 114:168–178

60. Aslanzadeh S, Rajendran K, Taherzadeh MJ (2014) A comparative study between single- and two-stage anaerobic digestion processes: effects of organic loading rate and hydraulic retention time. Int Biodeterior Biodegradation 95:181–188

61. Gou C, Yang Z, Huang J, Wang H, Xu H, Wang L (2014) Effects of temperature and organic loading rate on the performance and microbial community of anaerobic co-digestion of waste activated sludge and food waste. Chemosphere 105:146–151
62. Fu X, Hu Y (2016) Comparison of reactor configurations for biogas production from rapeseed straw. BioResources 11:9970–9985
63. Zarkadas IS, Sofikiti AS, Voudrias EA, Pilidis GA (2015) Thermophilic anaerobic digestion of pasteurised food wastes and dairy cattle manure in batch and large volume laboratory digesters: Focussing on mixing ratios. Renew Energy 80:432–440
64. Siddique MNI, Zularisam AW (2014) Intensified CSTR for bio-methane generation from petrochemical wastewater. In: International conference on chemical environment and biological science CEBS-2014, pp 17–18
65. Atelge MR, Krisa D, Kumar G, Eskicioglu C, Nguyen DD, Chang SW, Atabani AE, Al-Muhtaseb AH, Unalan S (2018) Biogas production from organic waste: recent progress and perspectives. Waste Biomass Valorization
66. Nagao N, Tajima N, Kawai M, Niwa C, Kurosawa N, Matsuyama T, Yusoff F, Toda T (2012) Maximum organic loading rate for the single-stage wet anaerobic digestion of food waste. Bioresour Technol 118:210–218
67. Wang L, Shen F, Yuan H, Zou D, Liu Y, Zhu B, Li X (2014) Anaerobic co-digestion of kitchen waste and fruit/vegetable waste: lab-scale and pilot-scale studies. Waste Manag 34:2627–2633
68. Sevilla-Espinosa S, Solórzano-Campo M, Bello-Mendoza R (2010) Performance of staged and non-staged up-flow anaerobic sludge bed (USSB and UASB) reactors treating low strength complex wastewater. Biodegradation 21:737–751
69. Fang C, Boe K (2011) Angelidaki I (2011) Biogas production from potato-juice, a by-product from potato-starch processing, in upflow anaerobic sludge blanket (UASB) and expanded granular sludge bed (EGSB) reactors. Bioresour Technol 102:5734–5741
70. Banu JR, Kaliappan S, Yeom IT (2007) Treatment of domestic wastewater using upflow anaerobic sludge blanket reactor. Int J Environ Sci Technol 4:363–370
71. Banu JR, Arulazhagan P, Kumar SA, Kaliappan S, Lakshmi AM (2015) Anaerobic co-digestion of chemical- and ozone-pretreated sludge in hybrid upflow anaerobic sludge blanket reactor. Desalin Water Treat 54:3269–3278
72. Goswami R, Chattopadhyay P, Shome A, Banerjee SN, Chakraborty AK, Mathew AK, Chaudhury S (2016) An overview of physico-chemical mechanisms of biogas production by microbial communities: a step towards sustainable waste management. 3 Biotech 6:72
73. Ziemiński K, Frąc M (2012) Methane fermentation process as anaerobic digestion of biomass: transformations, stages and microorganisms. African J Biotechnol 11:4127–4139
74. Kaparaju P, Serrano M, Angelidaki I (2010) Optimization of biogas production from wheat straw stillage in UASB reactor. Appl Energy 87:3779–3783
75. Meng Y, Jost C, Mumme J, Wang K, Linke B (2016) An analysis of single and two stage, mesophilic and thermophilic high-rate systems for anaerobic digestion of corn stalk. Chem Eng J 288:79–86
76. Böske J, Wirth B, Garlipp F, Mumme J, Van den Weghe H (2015) Upflow anaerobic solid-state (UASS) digestion of horse manure: thermophilic versus mesophilic performance. Bioresour Technol 175:8–16
77. Mumme J, Linke B, Tölle R (2010) Novel upflow anaerobic solid-state (UASS) reactor. Bioresour Technol 101:592–599
78. Pohl M, Mumme J, Heeg K, Nettmann E (2012) Thermo-and mesophilic anaerobic digestion of wheat straw by the upflow anaerobic solid-state (UASS) process. Bioresour Technol 124:321–327
79. Pohl M, Heeg K, Mumme J (2013) Anaerobic digestion of wheat straw–Performance of continuous solid-state digestion. Bioresour Technol 46:408–415
80. Böske J, Wirth B, Garlipp F, Mumme J, Van den Weghe H (2014) Anaerobic digestion of horse dung mixed with different bedding materials in an upflow solid-state (UASS) reactor at mesophilic conditions. Bioresour Technol 158:111–118

81. Haandel AV, Kato MT, Cavalcanti PFF, Florencio L (2006) Anaerobic reactor design concepts for the treatment of domestic wastewater. Rev Environ Sci Bio/Technol 5:21–38

82. Jafari J, Mesdaghinia A, Nabizadeh R, Farrokhi M, Mahvi AH (2013) Investigation of anaerobic fluidized bed reactor/aerobic moving bed bio reactor (AFBR/MMBR) system for treatment of currant wastewater. Iran J Public Health 42:860

83. Andalib M, Elbeshbishy E, Mustafa N, Hafez H, Nakhla G, Zhu J (2014) Performance of an anaerobic fluidized bed bioreactor (AnFBR) for digestion of primary municipal wastewater treatment biosolids and bioethanol thin stillage. Renew Energy 71:276–285

84. Omil F, Garrido JM, Arrojo B, Méndez R (2003) Anaerobic filter reactor performance for the treatment of complex dairy wastewater at industrial scale. Water Res 37:4099–4108

85. Mao C, Feng Y, Wang X, Ren G (2015) Review on research achievements of biogas from anaerobic digestion. Renew Sustain Energy Rev 45:540–555

86. Karunarathne HDSS (2015) Investigation of stability of plug flow anaerobic digester using mathematical modeling.

87. Perera U (2011) Investigation of operating conditions for optimum biogas production in plug flow type reactor

88. Hasangika WAS, Jaanuvi S, Karunathilake HP, Manthilake M, Rathnasiri PG, Sugathapala AGT (2015) Potential of plug flow digesters for biogas production in the Sri Lankan domestic context. In: Moratuwa Engineering Research Conference (MERCon). IEEE, pp 188–193

89. Jang HM, Ha JH, Kim M-S, Kim J-O, Kim YM, Park JM (2016) Effect of increased load of high-strength food wastewater in thermophilic and mesophilic anaerobic co-digestion of waste activated sludge on bacterial community structure. Water Res 99:140–148

90. Wang X, Li Z, Bai X, Zhou X, Cheng S, Gao R, Sun J (2018) Study on improving anaerobic co-digestion of cow manure and corn straw by fruit and vegetable waste: methane production and microbial community in CSTR process. Bioresour Technol 249:290–297

91. Campanaro S, Treu L, Kougias PG, De Francisci D, Valle G, Angelidaki I (2016) Metagenomic analysis and functional characterization of the biogas microbiome using high throughput shotgun sequencing and a novel binning strategy. Biotechnol Biofuels 9:26

92. Li L, Peng X, Wang X, Wu D (2018) Anaerobic digestion of food waste: a review focusing on process stability. Bioresour Technol 248:20–28

93. Li L, He Q, Ma Y, Wang X, Peng X (2016) A mesophilic anaerobic digester for treating food waste: process stability and microbial community analysis using pyrosequencing. Microb Cell Fact 15:65

94. Kim B-C, Kim S, Shin T, Kim H, Sang B-I (2016) Comparison of the bacterial communities in anaerobic, anoxic, and oxic chambers of a pilot A2O process using pyrosequencing analysis. Curr Microbiol 66:555–565

95. Lee HJ, Jung JY, Oh YK, Lee S-S, Madsen EL, Jeon CO (2012) Comparative survey of rumen microbial communities and metabolites across one caprine and three bovine groups, using bar-coded pyrosequencing and 1H nuclear magnetic resonance spectroscopy. Appl Environ Microbiol 78:5983–5993

96. Sundberg C, Al-Soud WA, Larsson M, Alm E, Yekta SS, Svensson BH, Sørensen SJ, Karlsson A (2013) 454 pyrosequencing analyses of bacterial and archaeal richness in 21 full-scale biogas digesters. FEMS Microbiol Ecol 85:612–626

97. Parameswaran P, Jalili R, Tao L, Shokralla S, Gharizadeh B, Ronaghi M, Fire AZ (2007) A pyrosequencing-tailored nucleotide barcode design unveils opportunities for large-scale sample multiplexing. Nucleic Acids Res 35:e130

98. Theuerl S, Kohrs F, Benndorf D, Maus I, Wibberg D, Schlüter A, Kausmann R, Heiermann M, Rapp E, Reichl U, Pühler A, Klocke M (2015) Community shifts in a well-operating agricultural biogas plant: how process variations are handled by the microbiome. Appl Microbiol Biotechnol 99:7791–7803

99. Luo G, De Francisci D, Kougias PG, Laura T, Zhu X, Angelidaki I (2015) New steady-state microbial community compositions and process performances in biogas reactors induced by temperature disturbances. Biotechnol Biofuels 8:3

100. Goux X, Calusinska M, Lemaigre S, Marynowska M, Klocke M, Udelhoven T, Benizri E, Delfosse P (2015) Microbial community dynamics in replicate anaerobic digesters exposed sequentially to increasing organic loading rate, acidosis, and process recovery. Biotechnol Biofuels 8:122

101. Jang HM, Kim JH, Ha JH, Park JM (2014) Bacterial and methanogenic archaeal communities during the single-stage anaerobic digestion of high-strength food wastewater. Bioresour Technol 165:174–182

102. Jang HM, Cho HU, Park SK, Ha JH, Park JM (2014) Influence of thermophilic aerobic digestion as a sludge pre-treatment and solids retention time of mesophilic anaerobic digestion on the methane production, sludge digestion and microbial communities in a sequential digestion process. Water Res 48:1–14

103. Li L, He Q, Ma Y, Wang X, Peng X (2015) Dynamics of microbial community in a mesophilic anaerobic digester treating food waste: relationship between community structure and process stability. Bioresour Technol 189:113–120

104. Carballa M, Smits M, Etchebehere C, Boon N, Verstraete W (2011) Correlations between molecular and operational parameters in continuous lab-scale anaerobic reactors. Appl Microbiol Biotechnol 89:303–314

105. Nakasaki K, Kwon SH, Takemoto Y (2015) An interesting correlation between methane production rates and archaea cell density during anaerobic digestion with increasing organic loading. Biomass Bioenerg 78:17–24

106. Lim JW, Chen C-L, Ho IJR, Wang J-Y (2013) Study of microbial community and biodegradation efficiency for single-and two-phase anaerobic co-digestion of brown water and food waste. Bioresour Technol 147:193–201

107. Nelson MC, Morrison M, Yu Z (2011) A meta-analysis of the microbial diversity observed in anaerobic digesters. Bioresour Technol 102:3730–3739

108. Wirth R, Kovács E, Maróti G, Bagi Z, Rákhely G, Kovács KL (2012) Characterization of a biogas-producing microbial community by short-read next generation DNA sequencing. Biotechnol Biofuels 5:41

109. Aydin S, Ince B, Ince O (2015) Application of real-time PCR to determination of combined effect of antibiotics on Bacteria, Methanogenic Archaea, Archaea in anaerobic sequencing batch reactors. Water Res 76:88–98

110. Chen S, Dong X (2005) Proteiniphilum acetatigenes gen. nov., sp. nov., from a UASB reactor treating brewery wastewater. Int J Syst Evol Microbiol 55:2257–2261

111. Wang P, Wang H, Qiu Y, Ren L, Jiang B (2018) Microbial characteristics in anaerobic digestion process of food waste for methane production—a review. Bioresour Technol 248:29–36

112. Rincón B, Portillo MC, González JM, Borja R (2013) Microbial community dynamics in the two-stage anaerobic digestion process of two-phase olive mill residue. Int J Environ Sci Technol 10:635–644

113. Merlino G, Rizzi A, Schievano A, Tenca A, Scaglia B, Oberti R, Adani F, Daffonchio D (2013) Microbial community structure and dynamics in two-stage vs single-stage thermophilic anaerobic digestion of mixed swine slurry and market bio-waste. Water Res 47:1983–1995

114. Ziganshin AM, Liebetrau J, Pröter J, Kleinsteuber S (2013) Microbial community structure and dynamics during anaerobic digestion of various agricultural waste materials. Appl Microbiol Biotechnol 97:5161–5174

115. Motte J-C, Trably E, Escudié R, Hamelin J, Steyer J-P, Bernet N, Delgenes J-P, Dumas C (2013) Total solids content: a key parameter of metabolic pathways in dry anaerobic digestion. Biotechnol Biofuels 6:164

116. Lindmark J, Thorin E, Bel Fdhila R, Dahlquist E (2014) Effects of mixing on the result of anaerobic digestion: review. Renew Sustain Energy Rev 40:1030–1047

117. Kundu K, Sharma S, Sreekrishnan TR (2012) Effect of operating temperatures on the microbial community profiles in a high cell density hybrid anaerobic bioreactor. Bioresour Technol 118:502–511

118. Mamimin C, Singkhala A, Kongjan P, Suraraksa B, Prasertsan P, Imai T, O-Thong S, (2015) Two-stage thermophilic fermentation and mesophilic methanogen process for biohythane production from palm oil mill effluent. Int J Hydrogen Energy 40:6319–6328

119. hamed A, Chen C-L, Rajagopal R, Wu D, Mao Y, Ho IJR, Lim JW, Wang JY, (2015) Multi-phased anaerobic baffled reactor treating food waste. Bioresour Technol 182:239–244
120. Álvarez JA, Armstrong E, Gómez M, Soto M (2008) Anaerobic treatment of low-strength municipal wastewater by a two-stage pilot plant under psychrophilic conditions. Bioresour Technol 99:7051–7062
121. Wu L-J, Kobayashi T, Li Y-Y, Xu K-Q (2015) Comparison of single-stage and temperature-phased two-stage anaerobic digestion of oily food waste. Energy Convers Manag 106:1174–1182
122. Wang TX, Ma XY, Wang MM, Chu HJ, Zuo JE, Yang YF (2016) A comparative study of micro-bial community compositions in thermophilic and mesophilic sludge anaerobic digestion systems. Microbiol China 43:26–35
123. Wu L-J, Higashimori A, Qin Y, Hojo T, Kubota K, Li Y-Y (2016) Comparison of hyper-thermophilic–mesophilic two-stage with single-stage mesophilic anaerobic digestion of waste activated sludge: process performance and microbial community analysis. Chem Eng J 290:290–301
124. Shimada T, Morgenroth E, Tandukar M, Pavlostathis SG, Smith A, Raskin L, Kilian RE (2011) Syntrophic acetate oxidation in two-phase (acid–methane) anaerobic digesters. Water Sci Technol 64:1812–1820
125. Maspolim Y, Zhou Y, Guo C, Xiao K, Ng WJ (2015) Determination of the archaeal and bacte-rial communities in two-phase and single-stage anaerobic systems by 454 pyrosequencing. J Environ Sci 36:121–129
126. Venkiteshwaran K, Bocher B, Maki J, Zitomer D (2015) relating anaerobic digestion micro-bial community and process function: supplementary issue: water microbiology. Microbiol Insights 8:MBI-S33593
127. Liu C, Li H, Zhang Y, Si D, Chen Q (2016) Evolution of microbial community along with increasing solid concentration during high-solids anaerobic digestion of sewage sludge. Bioresour Technol 216:87–94
128. Guo J, Peng Y, Ni B-J, Han X, Fan L, Yuan Z (2015) Dissecting microbial community structure and methane-producing pathways of a full-scale anaerobic reactor digesting activated sludge from wastewater treatment by metagenomic sequencing. Microb Cell Fact 14:33
129. Zhang L, Ban Q, Li J, Jha AK (2016) Response of syntrophic propionate degradation to pH decrease and microbial community shifts in an UASB reactor. J Microbiol Biotechnol 26:1409–1419
130. Leng L, Yang P, Singh S, Zhuang H, Xu L, Chen W-H, Dolfing J, Li D, Zhang Y, Zeng H, Chu W, Lee PH (2017) A review on the bioenergetics of anaerobic microbial metabolism close to the thermodynamic limits and its implications for digestion applications. Bioresour Technol 2017:1095–1106
131. Tang Y-Q, Shigematsu T, Morimura S, Kida K (2015) Dynamics of the microbial community during continuous methane fermentation in continuously stirred tank reactors. J Biosci Bioeng 119:375–383
132. Cai M, Wilkins D, Chen J, Ng S-K, Lu H, Jia Y, Lee PK (2016) Metagenomic reconstruction of key anaerobic digestion pathways in municipal sludge and industrial wastewater biogas-producing systems. Front Microbiol 7:778
133. Ho DP, Jensen PD, Batstone DJ (2013) Methanosarcinaceae and acetate oxidising path-ways dominate in high-rate thermophilic anaerobic digestion of waste activated sludge. Appl Environ Microbiol 2013:AEM-01730
134. Westerholm M, Dolfing J, Sherry A, Gray ND, Head IM, Schnürer A (2011) Quantification of syntrophic acetate-oxidizing microbial communities in biogas processes. Environ Microbiol Rep 3:500–505
135. Luo G, Fotidis IA, Angelidaki I (2016) Comparative analysis of taxonomic, functional, and metabolic patterns of microbiomes from 14 full-scale biogas reactors by metagenomic sequencing and radioisotopic analysis. Biotechnol Biofuels 9:51
136. Zhang L, Lee Y-W, Jahng D (2011) Anaerobic co-digestion of food waste and piggery wastewater: focusing on the role of trace elements. Bioresour Technol 102:5048–5059

137. He Y, Pang Y, Liu Y, Li X, Wang K (2008) Physicochemical characterization of rice straw pretreated with sodium hydroxide in the solid state for enhancing biogas production. Energy Fuels 22:2775–2781

138. Kariyama ID, Zhai X, Wu B (2018) Influence of mixing on anaerobic digestion efficiency in stirred tank digesters: a review. Water Res 143:503–517

139. De Baere L, Mattheeuws B, Velghe F (2010) State of the art of anaerobic digestion in Europe. In: 12th world congress on anaerobic digestion, Guadalajara Mexico, pp 3–6

140. De Gioannis G, Diaz LF, Muntoni A, Pisanu A (2008) Two-phase anaerobic digestion within a solid waste/wastewater integrated management system. Waste Manag 28:1801–1808

141. Shanmugam AS, Akunna JC (2010) Modelling head losses in granular bed anaerobic baffled reactors at high flows during start-up. Water Res 44:5474–5480

142. Ahamed A, Chen C-L, Rajagopal R, Wu D, Mao Y, Ho IJR, Lim JW, Wang JY (2015) Multi-phased anaerobic baffled reactor treating food waste. Bioresour Technol 182:239–244

143. Castellano-Hinojosa A, Armato C, Pozo C, González-Martínez A, González-López J (2018) New concepts in anaerobic digestion processes: recent advances and biological aspects. Appl Microbiol Biotechnol 102:5065–5076

144. Xie S, Hai FI, Zhan X, Guo W, Ngo HH, Price WE, Nghiem LD (2016) Anaerobic co-digestion: A critical review of mathematical modelling for performance optimization. Bioresour Technol 222:498–512

145. Abbasi T, Ramasamy E V, Khan FI, Abbasi SA (2012) Regional EIA and risk assessment in a fast developing country

146. Battista F, Fino D, Mancini G (2016) Optimization of biogas production from coffee production waste. Bioresour Technol 200:884–890

147. Zahedi S, Sales D, Romero LI, Solera R (2013) Optimisation of single-phase dry-thermophilic anaerobic digestion under high organic loading rates of industrial municipal solid waste: Population dynamics. Bioresour Technol 146:109–117

148. Pöschl M, Ward S, Owende P (2010) Evaluation of energy efficiency of various biogas production and utilization pathways. Appl Energy 87:3305–3321

149. Arsova L (2010) Anaerobic digestion of food waste: current status, problems and an alternative product

150. Ferrer I, Garfí M, Uggetti E, Ferrer-Martí L, Calderon A, Velo E (2011) Biogas production in low-cost household digesters at the Peruvian Andes. Biomass Bioenerg 35:1668–1674

151. Saravanan SS, Vivekanandan S (2014) Effect of mesosphilic and thermophilic temperature on floating drum anaerobic bio-digester, vol 14

152. Ponsá S, Ferrer I, Vázquez F, Font X (2008) Optimization of the hydrolytic–acidogenic anaerobic digestion stage (55 C) of sewage sludge: Influence of pH and solid content. Water Res 42:3972–3980

153. Ferrer I, Serrano E, Ponsa S, Vazquez F, Font X (2009) Enhancement of thermophilic anaerobic sludge digestion by 70 °C pre-treatment: energy considerations. J Residuals Sci Technol 6:11–18

154. Choorit W, Wisarnwan P (2007) Effect of temperature on the anaerobic digestion of palm oil mill effluent. Electron J Biotechnol 10:376–385

155. Aboudi K, Quiroga XG, Gallego CJA, Gallego LIR (2017) Comparison of single-stage and temperature-phased anaerobic digestion of sugar beet by-products. In: 5th international conference on sustainable solid waste management, 24–27 June, pp 1–24

156. Yu L, Wensel PC, Ma JW, Chen SL (2014) Mathematical modeling in anaerobic digestion (AD). J Bioremediation Biodegrad 5:003

157. Jha AK, Li J, Nies L, Zhang L (2011) Research advances in dry anaerobic digestion process of solid organic wastes. Afr J Biotechnol 10:14242–14253

158. Adekunle KF, Okolie JA (2015) A review of biochemical process of anaerobic digestion. Adv Biosci Biotechnol 6:205

159. Prasad RD (2012) Empirical study on factors affecting biogas production. ISRN Renew Energy 2012:136959

160. Chen X, Romano RT, Zhang R (2010) Anaerobic digestion of food wastes for biogas production. Int J Agric Biol Eng 3:61–72
161. Cruz-Salomón A, Ríos-Valdovinos E, Pola-Albores F, Meza-Gordillo R, Lagunas-Rivera S, Ruíz-Valdiviezo VM (2017) Anaerobic treatment of agro-industrial wastewater for COD removal in expanded granular sludge bed bioreactor. Biofuel Res J 16:715–720
162. Lindmark J, Eriksson P, Thorin E (2014) The effects of different mixing intensities during anaerobic digestion of the organic fraction of municipal solid waste. Waste Manag 34:1391–1397
163. Fagbohungbe MO, Dodd IC, Herbert BMJ, Li H, Ricketts L, Semple KT (2015) High solid anaerobic digestion: operational challenges and possibilities. Environ Technol Innov 4:268–284
164. Fdéz.-Güelfo LA, Álvarez-Gallego C, Sales Márquez D, Romero García LI, (2010) Start-up of thermophilic–dry anaerobic digestion of OFMSW using adapted modified SEBAC inoculum. Bioresour Technol 101:9031–9039
165. Cho S-K, Im W-T, Kim D-H, Kim M-H, Shin H-S, Oh S-E (2013) Dry anaerobic digestion of food waste under mesophilic conditions: performance and methanogenic community analysis. Bioresour Technol 131:210–217
166. Kress P, Naegele H-J, Oechsner H, Ruile S (2018) Effect of agitation time on nutrient distribution in full-scale CSTR biogas digesters. Bioresour Technol 247:1–6
167. Afridi ZUR, Wu J, Cao ZP, Zhang ZL, Li ZH, Poncin S, Li HZ (2017) Insight into mass transfer by convective diffusion in anaerobic granules to enhance biogas production. Biochem Eng J 127:154–160
168. De Jesus SS, Neto JM, Maciel Filho R (2017) Hydrodynamics and mass transfer in bubble column, conventional airlift, stirred airlift and stirred tank bioreactors, using viscous fluid: a comparative study. Biochem Eng J 118:70–81
169. Rajagopal R (2008) Treatment of agro-food industrial wastewaters Using UAF and Hybrid UASB-UAF reactors
170. Feng Y, Lu B, Jiang Y, Chen Y, Shen S (2012) Anaerobic degradation of purified terephthalic acid wastewater using a novel, rapid mass-transfer circulating fluidized bed. Water Sci Technol 65:1988–1993
171. Li W-W, Yu H-Q (2011) Physicochemical characteristics of anaerobic H2-producing granular sludge. Bioresour Technol 102:8653–8660
172. Shahid MK, Kashif A, Rout PR, Aslam M, Fuwad A, Choi Y, Banu JR, Park JH, Kumar G (2020) A brief review of anaerobic membrane bioreactors emphasizing recent advancements, fouling issues and future perspectives. J Environ Manage 270:110909
173. Brown D, Li Y (2013) Solid state anaerobic co-digestion of yard waste and food waste for biogas production. Bioresour Technol 127:275–280
174. Khalid A, Arshad M, Anjum M, Mahmood T, Dawson L (2011) The anaerobic digestion of solid organic waste. Waste Manag 31:1737–1744
175. Rapport JL, Zhang R, Williams RB, Jenkins BM (2012) Anaerobic digestion technologies for the treatment of municipal solid waste. Int J Environ Waste Manag 9:100–122
176. Paranjpe A, Sharma AK, Ranjan RK, Paranjape, (2013) MSW A potential energy resources: a two stage anaerobic digestio. Asian J Sci Appl Technol 2:1–7
177. Sakai S, Tsuchida Y, Okino S, Ichihashi O, Kawaguchi H, Watanabe T, Inui M, Yukawa H (2007) Effect of lignocellulose-derived inhibitors on growth of and ethanol production by growth-arrested Corynebacterium glutamicum R. Appl Environ Microbiol 73:2349–2353
178. Lv W, Schanbacher FL, Yu Z (2010) Putting microbes to work in sequence: recent advances in temperature-phased anaerobic digestion processes. Bioresour Technol 101:9409–9414
179. Gourdet C, Girault R, Berthault S, Richard M, Tosoni J, Pradel M (2017) In quest of environmental hotspots of sewage sludge treatment combining anaerobic digestion and mechanical dewatering: a life cycle assessment approach. J Clean Prod 143:1123–1136

Chapter 9
Two-Stage Anaerobic Digest

9.1 Two-Stage Digestion

Two-stage anaerobic digestions (2S-AD) is one such innovation which embodies a acidogenesis (stage 1) and a methanogenesis (stage 2) [1]. 2S-AD leverages greater stable performance and higher yields [2] and in addition to methane, it can also be used for producing hydrogen at the acidogenesis stage [3]. Contrasting with the 1S-systems, 2S-systems aid AD owing to the buffering impact of substrate loading in stage 1 and provide a suitable condition in stage 2. Thus, these benefits advocate proficiency with regard to substrate biodegradation and bioenergy generation [4].

9.2 Two-Stage Digester Configurations

In 2S-AD, the physical partition of the reactors which incorporates both stage 1 and stage 2 empowers a suitable environment for microbial population from both stages (acidogenesis and methanogenesis) with improved growth conditions and sequentially exaggerates biogas production [5]. Researchers customarily warranted that 2S-AD are endorsed to conventional 1S-AD concerning to biogas generation and stable digestion performance [6]. The main aspects to enhance the organic load, biogas productivity and stable performance of the digestion system are improvement in bioreactors configuration and operating strategies. The problems faced by digesters are rapid acidification, inhibition of methanogens, etc. To overcome these limitations, the two-stage digesters have been developed. In these systems, the production of acid and methane are parted into two digesters to overcome the issues of pH inhibition faced by single stage digestion [7]. In two-stage digesters, the acidic pH ranging from 5.5 to 6.5 and retention time of 2–3 days was normally maintained in stage 1 acidogenic digesters. In stage 2 methanogenic digesters, the pH in the range of 6–8 and HRT of 20–30 days were maintained to proliferate the slow growth of methanogens [8]. To improve the substrate biodegradation and bioenergy recovery,

K. Sudalyandi and R. Jeyakumar, *Biofuel Production Using Anaerobic Digestion*,
Green Energy and Technology, https://doi.org/10.1007/978-981-19-3743-9_9

two-stage digestion is an excellent choice where the first digester is used for production of biohydrogen. The effluent generated from first digester is recycled into the second digester as substrate to produce methane. Two-stage digestion have been developed and executed over 20 years. At present many design, modifications and improvements have been configured in two-stage systems to enhance its efficiency. Various forms of two-stage reactors fostered at present are catalogued in the following section.

Integrated reactors

The two-stage digestion systems include integrated systems of anaerobic filter–UASB (AF–UASB) reactor, anaerobic solid–liquid (HASL)–UASB system, Anaerobic UASB–AFBR reactors, CSTR–CSTR, CSTR–ABR (anaerobic baffled reactor), CSTR–UASB, UASB–UASB, ASBR–UASB and Stepped anaerobic baffled reactor (SABR), Reactors with phase separation, Temperature phased anaerobic reactors and Two phased CSTR–PBAR (pressurized biofilm anaerobic reactor).

Integrated anaerobic filter–UASB (AF–UASB) reactor

Microbial population washout is the main issue faced in UASB for treating wastewater generating from industries. AF–UASB was designed by creating modification in gas liquid separator. This is an integrated reactor in which the anaerobic filter was located above the UASB. Therefore, the performance benefits of both the AF and USAB were integrated in these types of reactors. The suspension of microbial development occurs at the base of the reactor. This serves as the separating area for the separation of contaminated and recalcitrant substances exist in substrate. The contaminated and recalcitrant substances include fat, glycerides, phenolics and other non-biodegradable materials. The top portion of these reactors have bed of filters that manage organic rich influent and let the biomass to get attached to the bed [9]. These types of reactors could be employed for treating wastewaters generating from various industries such as chemical industries, distillery effluents and effluents rich in phenolic compounds. The AF–UASB reactor performance of this approach in treating palm industry wastewater, show that a COD removal efficiency up to 98% and it also have shortened the startup period from 60 days as in UASB reactors to 47 days to achieve steady state [10]. The treatment efficiency of UASB and anaerobic integrated reactor for the removal of phenolics at different HRTs (0.75–0.33 d) has investigated by a study [11]. Within the startup and microbial granule formation in anaerobic integrated reactor was found rapid. Conjointly, reduction in retention time caused drop in biodegradation of phenolic compounds in anaerobic integrated reactor (99–77%) and UASB (95–68%) besides withstanding a greater shock load than UASB.

Integrated anaerobic solid–liquid (IASL)–UASB system

This phase separated reactor connects two reactors (acidogenic and methanogenic) as shown in Fig. 9.1a. In these types of reactors, the substrates (wastes in solid form) are loaded into the acidogenic reactor. The liquid which bobs up are recycled to the UASB for methane generation [12]. An investigation of the recirculating impact of food waste leachate produced in the acidogenic reactor during digestion process in integrated anaerobic solid–liquid reactor has been evaluated by a study [13]. This recirculating effect of leachate receded the time required for producing equivalent similar aggregate of methane by 40% when compared to without recirculation.

Anaerobic UASB–AFBR integrated reactors

This type of integrated reactors possesses artificial liquid velocity superior to USAB and subordinate to AFBR. These reactors were designed and employed in a study [14]. These reactors allows the fluidization of granular sludge biomass, thus capping the traits conjoint to both UASB and AFBR (Fig. 8.4c). At a substrate loading of $2.08 \, kg \, COD \, m^{-3} \, d^{-1}$ and retention time of 6 h, these reactors removes approximately 94% organics. It is reported that in an integrated reactor, the operational temperature of 37 °C was observed as optimal for effective functioning and for efficient existence of various microbial populations [15]. The genomic study of microbial population in the integrated reactors working at elevated temperature of 45–55 °C shown only the minor existence of diverse methanogenic microbes. This, palpably decline the potency of organics removal in the reactors in comparison to 37 °C temperature.

CSTR–CSTR system

CSTR is an appropriate option for the stage 1, while the choice of the stage 2 digester reliant on characteristics of feed supplied from the stage 1 digester. CSTR can be apt for all kind of feed irrespective of its characteristics. The two-stage anaerobic digestion with CSTR in both stages, can capable of governing the acidification process in the stage 1 and regulation of volatile fatty acids (VFAs) in the stage 2. Thus, a steady accomplishment in process was achieved by integrating CSTR with CSTR. A two-stage integrated mesophilic CSTRs were operated in a study [16] for hydrogen and methane generation from cheese whey wastewater at an operating substrate load of 60 g/L. A HRT for 24 h was maintained in first stage CSTR producing hydrogen and a HRT of 20 days was maintained in second stage CSTR producing methane. Sugar utilization in the first stage digester was nearly 86% and the production of hydrogen was found to be $0.052 \, m^3 \, H_2/kg \, COD$. Together both stages were responsible for a COD removal efficiency of 95.3%. A two-stage CSTR–CSTR configuration used in a study [17] for biohydrogen and biomethane production. Where they fed the digesters with synthetic medium containing glucose and used anaerobically digested sludge as inoculums. A methane and hydrogen yield of 2.75 and 2.13 mol/mol hexose were obtained at an optimized inoculums-to-substrate ratio of 2:1 with an energy recovery of 82%. The same two-stage CSTR–CSTR configuration, was employed by two different study [18, 19] for the production of biofuel. They stated that installation of gravity settler next to hydrogen producing CSTR stabilized the process by retaining hydrogen producing microbes in the digester.

CSTR–ABR system

The integration of CSTR and ABR can be a suitable option to overcome the drawbacks of CSTR and ABR digesters. High-rate anaerobic reactors are typically incapable to hold the larger solid particles and lipid contents present in substrates. This can

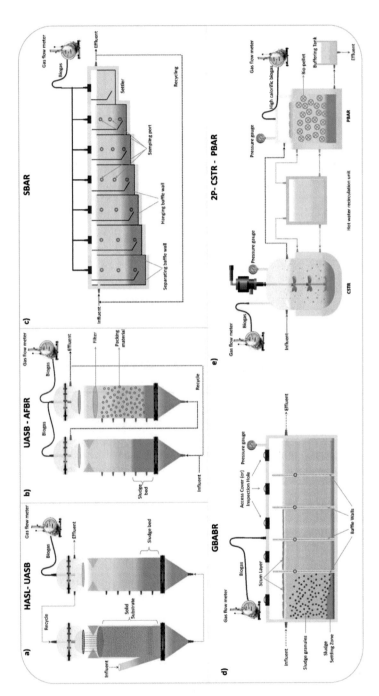

Fig. 9.1 Two-stage digester configurations

be averted in high rate two-stage reactors by coupling CSTR. The integrated CSTR and ABR systems is beneficial as they complement with each other to overcome their individual drawbacks. A two-stage digester, with CSTR and periodic anaerobic baffled reactor (PABR) was used to treat whey wastewater at mesophilic condition [20]. The retention time for CSTR was 24 h and PABR was 20, 10 and 4.4 days. The organic removal efficiency of CSTR was reported as lower than 5% and the hydrogen yield was found to be 0.041 m^3 per kg COD. The lower yield of hydrogen yield in the CSTR implied that the produced hydrogen was utilized by the homoacetogens and some organics were transformed to acetic and butyric acid. In PABR at 20 and 10 day HRT, the COD removal efficiency was nearly 99%. The removal efficiency was recorded as 94% at 4.4 days HRT. The methane contend of biogas was reported to be greater than 70% in all operating conditions.

CSTR–UASB system

Incorporation of UASB in second stage digestion offers several benefits. Conventionally low rate digesters were employed in second stage. These digesters normally possess extended retention times. This shortcoming can be overcome with a high rate UASB that permits short retention times. In UASB biomass flocculate and form granules and the granulated biomass have good settling property. The shortcoming of UASB is that this technology cannot handle high solid content [21]. A two-stage mesophilic CSTR–UASB was configured [22] for treating whey powder provided with micronutrients. In CSTR, hydrogen production was approximately 50% with no methane generation. The treated wastewater of CSTR was effectively biodegraded in the UASB with substrate biodegradation efficiency of 90%. The hydrogen production was recorded as 0.156 m^3 H$_2$/kg COD at 6 h HRT. The same study reported that the hydrogen production at 3.5 h HRT, but the production seems to be not stable. In this study 70% increment in energy recovery have been obtained when the CSTR was integrated with UASB reactor. A two-stage CSTR (thermophilic)–UASB (mesophilic) configuration was employed in a study [23] to generate energy from the wastewater discharged generating from palm oil industry. In this configuration, recirculating methanogenic effluent to CSTR, significantly influenced biofuel production. When, methanogenic effluent was recirculated into CSTR at a rate of 30%, it kept pH in balance, which resulted in double the time increment in H$_2$ generation. The H$_2$ and CH$_4$ production have been reported to be 135 and 414 mL CH$_4$/g VS. The biohythane consist of 13.3% hydrogen, 54.4% methane, and 32.2% carbon dioxide. Besides, recirculating treated wastewater from methanogenic reactor supports the development and activity of *Thermoanaerobacterium* sp. in biohydrogen digester as it enhance biohydrogen production.

UASB–UASB system

The two-stage system integrating UASB–UASB was used to generate energy from wastewaters. UASB was designed to retain maximum amount of biomass and it forms a blanket at the bottom of the reactor. Biomass of UASB is granulated and having good settling property. Owing to retention of massive biomass concentration, UASB produce higher amount of biofuel and wastewater treatment efficiency than others.

A two-stage sequential mesophilic UASB digesters were run for treating whey at a pH of 5.8 for hydrogen and at 7.4 for methane production [24]. An increment in substrate loading from 20 to 30 kg COD/m^3/d increased hydrogen generation from 0.0167 to 0.035 m^3 H_2/kg COD. Further increase in organic loading deteriorated UASB performance. In another study [25] a two-stage digestion system was used which was comprising of extreme thermophilic and thermophilic UASB for treating wheat straw hydrolysate. The energy recoveries from these two digesters were reported to be 89 mL-H_2/g-VS and 307 mL CH_4/g VS respectively. A volatile solids removal of 81% was obtained by this integrated reactors. The composition of biohythane consist of 16.5% hydrogen, 44.8% methane, and 38.7% carbon dioxide. Microbial populations such as *Thermoanaerobacter wiegelii*, and *Caloramatorfervidus* were accountable for generation of hydrogen in first stage UASB. The predominant microbial population accountable for methane production in second stage UASB were *Methanosarcina mazei* and *Methanothermobacter defluvii*.

ASBR–UASB system

The two-stage ASBR–UASB can improve the biodegradation of substrate and can overcome the negative effects of inhibitory components in the substrate [26]. In addition, two-stage ASBR–UASB system have benefits of improved the net energy production, enhanced stability with easier regulation of acidogenic stage, greater organic loading capability and improved activity of specific methanogenic populations. This causes an increase in methane production and enhanced substrate removal efficiency, when compared to 1S-AD. An efficient energy recovery from palm oil industry wastewater was achieved in a study [27] through two-stage ASBR (thermophilic)–UASB (mesophilic) digesters. This system produced greater biohythane production (4.4 L/L d) with it contain 14% hydrogen, 51% methane and 35% carbon dioxide.

Stepped anaerobic baffled reactor (SABR)

The stepped anaerobic baffled reactor (SABR) was considered to be one of the advanced and newly designed digester for the production of biohythane in an experimental study [28] conducted for the generation of biohythane from petrochemical industry effluents rich in mono-ethylene glycol as a substrate. It run for 5 months with a HRT of 72 h. The achieved hydrogen and methane yield was found to be 88 mL H_2/g volatile solids and 318 mL CH_4/g volatile solids, respectively. Figure 9.1c shows a schematic representation of SABR.

Reactors with Phase Separation

The UASB and EGSB are the basic reactors from which the phase-separation reactors branch out to integrated digesters.

Upflow Staged Sludge Bed (USSB) Reactor

The upflow anaerobic solid-state (UASS) reactor treats wastes (solid form) whereas the complex materials arise as bed (in the form of solid). This reactor configuration was contemplated in a study [29]. In these reactors, the complex materials move vertically and is enticed via the suture of microaerobically generated bubbles. This reactor is separated into triple phases: an underneath liquefied phase, a top liquefied phase and the bed of solids in the centre. The top termination of solid bed is connected with a filter at the upper portion of the reactor. This filter acts as a three-zone sieve and holds the solid bed under the liquid phase [30]. The removal of solids is set at the peak phase under the filter. Through liquid recycling, bacterial population is shifted to reactor underneath. To decrease the content of organic acids, the liquid recycling could be fitted with a biogas bioreactor. A study [31] in which modified UASS reactor as defined by Mumme et al. [29] probed the technical feasibility of corn stalk with an additional AF reactor collated 1S and 2S digesters. The 2S-AD reactors are highly stable and outturns higher biogas yield when compared to 1S-AD. [32].

Granular bed anaerobic baffled reactor (GBABR)

GBABR (Fig. 9.1d) mixes the 'pro' that leads to concurrence by means of granular sludge that is acting as barrier in UASB. This function as phase separated reactors and resembles the anaerobic baffled reactor [33]. This reactor is a vertical shaped vessel with interior perpendicular spargers dangling and upended interchangeably signifies the fundamental configuration. The spargers are fitted as mobile baffles. In these reactors, the sludge generating from acidogenic reactors settles at the extremely viable methanogenic granules phase. The methanogenic granular phase averts the torpedo of acidogenic reactor and alters the stable performance of the system. Also, the screening of inoculum and acclimating in acidogenic reactors are higher in the partition close to influent port whereas the methane forming granules are higher in the partition close to effluent port. The performance of the reactor with food waste (FW) was evaluated in a study [34] and it shows an effective biogas yield of 215.57 mL/g-VS removed/d. GBABR have a functional issue induced via the existence of back pressure that interrupts the inflow design and minimizes the reaction time of influent and microbes. This issue could be reduced through suitable placing of spargers within the digester [33].

Hydrolytic upflow anaerobic sludge blanket (HUSB) reactor

The addition of untreated municipal wastewater with elevated solids content in traditional UASB may cause descent in growth of methane producing microbes and sludge deforming and therefore setting of sludge is essential. Therefore, this condition is sequestered in 2S-AD where a sequence type hydrolysis-acidogenic and methane digester is employed Fig. 8.4a. The first reactor (HUASB) works under hydrolysis stage and in this reactor, the complex molecules are hydrolyzed into simpler ones. The generated organic acids are transferred to second reactor (methanogenesis UASB) instigating organics removal and methane yield [12]. This integrated reactor removes 50–100% of waste content and obtains a stability at mesophilic condition with a substrate biodegradation of 50–60% and solids reduction of 65–85% [35].

New modifications and configurations of two-stage reactors

Many advanced two-stage digesters are designed and configured for both hydrogen and methane production. Reactors must be designed to be practical and economical for bioenergy recovery. In this context, various digesters have been developed by researchers with a goal to achieve greater biofuel production.

Temperature phased two-stage anaerobic digesters

Temperature phased anaerobic digesters have been considered to be one of the modified two-stage digestion system. In these digesters, the temperature of the two-stages varies. A study [36] in which temperature phased two-stage digestion system by maintaining thermophilic condition in first stage followed by mesophilic condition in second stage for treating oily food wastes. The oily food contains higher fat or lipid content. Maintaining thermophilic temperatures in the first stage of the digestion system was expected to improve hydrolysis of lipids. Second stage reactor was operated with mesophilic temperature as it favors the growth of methanogenic population and enhances methane production. Maintaining thermophilic condition also enhances the pasteurization of feedstock. Effluent recirculation from stage II to stage I digester helps in maintaining the pH in excess of 5.0 or else due to acidification pH drops below 4 in stage I reactor and it hampers hydrolysis. The output energy obtained in the these type of digester (20.4 kJ/g $VS_{reduced}$) was comparable with the mesophilic digestion, however an extra energy of 3.4 kJ/g VS reduced was essential to uphold the temperature in this system and for wastewater recirculation.

Two phased CSTR–PBAR (2P-CSTR–PBAR)

Two phase CSTR–PBAR (pressurized biofilm anaerobic reactor) configuration is reported to be suitable for digesting food waste. Hydrogen sulphide accumulation was the main drawback that impacts the digester performance especially methanogenesis stage. The two-phase CSTR–PBAR digester can decrease the hydrogen sulphide and carbon dioxide content during methanogenesis. Figure 9.1e shows a schematic representation of 2P-CSTR–PBAR. A two-stage CSTR–PBAR designed for digesting food waste [37]. This two-stage system increased the content of methane from 80 to 90% upon increasing the PBAR pressure from 0.3 to 1.7 MPa. But, the increment in pressure in addition decreased the production of methane and substrate degradation efficiency. A sulfidogenesis process (reduction of sulphate to hydrogen sulphide) in acidogenic phase [38]. This was achieved by cultivating sulphate reducing microbes in the first-stage digester and it nearly reduced 78% of sulphate. As a result, hydrogen sulphide concentration in the second stage digester was considerably decreased to 200 ± 15 ppm. Whereas, its concentration was found to be much higher (1,650 ± 25 ppm) in 1S-AD.

9.2.1 Microbial Communities and Pathways

Microbial communities are the central workers of an anaerobic digestion process. Two-stage digestion systems distinctly divide the entire digestion into two phases such as, acidogenic and methanogenic. In two-stage systems, the microbial communities in the two phases varied considerably from each other [39]. Evaluation of microbial community structures and its abundant variations or alterations in two-stage digestion could deliver possible outcomes of reactor performance and stability. These outcomes can be investigated to figure out the favorable conditions in staged reactors to enhance the effective biodegradation of organic matter.

Microbial communities abundance and its changes

Acidogenic and methanogenic microbes are reported be the main microbial populations that imparts a major role in anaerobic digestion. Many reports have documented about the structure of microbial communities abundant in two-stage reactors that might vary depending upon the substrate and operating conditions [40]. Figure 9.2 shows a schematic representation of different microbial communities in two-stage AD. Table 9.1 represents the summary of microbial communities in various two-stage digesters at varying operational conditions. A study [40] investigated the potential of two-stage laboratory scale reactors treating cattle manure at an substrate loading of 6.25 g VS/L day; temp of 37 °C and retention time of 5 days (acidogenic) and 15 days (methanogenic). Upon analyzing the microbial community structure using Ion torrent sequencing in these reactors reported that they have *Firmicutes, Actinobacteria Bacteroidetes* are the major microbial communities. Among them, *Firmicutes* were reported to be higher in acidogenic bioreactor whereas *Bacteroidetes* and Actinobacteria were reported to be higher in methanogenic reactor. A two-stage CSTR for codigesting swine manure, fruits and vegetable waste at thermophilic condition (55 °C) with 3 days of HRT for acidogenic reactor and 22 days of HRT for methanogenic reactor [41]. During codigestion, they have assessed microbial communities through DGGE-real time quantitative PCR analysis and reported phylum *Firmicutes* existed predominantly with nine species that belongs to the order, *Clostridiales* and with two species that comes under the order, *Thermoanaerobacterales.* In addition, they reported about the abundances of thermophilic cellulose secreting bacterium, *Clostridium cellulosi* in acidogenic reactor. In another study, that have investigated microbial communities in two-stage mesophilic digester treating waste activated sludge [42]. They have analyzed the microbial communities through 16S rRNA sequencing—DGGE analysis and reported the predominance of phylum that includes *Firmicutes, Chloroflexi* (Phyla level); archaea that includes *Methanosaeta* in methanogenic reactor. A two-stage hyperthermophilic reactor treating waste activated sludge was employed to find the predominant microbial community using 16S rRNA cloning and sequencing [43]. It was reported that about 27.9% of microbial community belongs to phylum *Firmicutes* and 34.4% belongs to *Proteobacteria.* Many researchers have reported that alteration or variation in process parameters would lead to changes in existing microbial communities in the two-stage digester [44–46]. For example, Shin et al. [47] studied about the changes in microbial community structures with respect to variation in HRT during 2S-AD of organic waste. They have reported that the stability of acidogenic reactor get altered when the HRT is varied. The decrement in HRT inflict selective pressure on microbes which are slow growers and this could cause shifts in microbial community. The authors have noticed changes in microbial community structure by decreasing the HRT from 25 to 4 days. The dominance of acetate producing acidogenic bacterial species related to *Aeriscardovia aeriphila and Lactobacillus amylovorus* shifted to homofermentative lactate producing bacterial species, *Lactobacillus acetotolerans* and *Lactobacillus kefirilike*. Similarly, a drastic changes in microbial community structure also noted in methanogenic digesters with respect to decrement in HRT. The authors have noted a shift of bacterial species from *Methanoculleus* to *Methanosarcina* communities. This shift in microbial communities was analyzed by non-metric multidimensional scaling joint plot analysis. A study [48] investigated about the shift of archaeal communities (shift in hydrogenotrophic methanogens) in two-stage CSTR digesters treating food waste with respect to operational time period. It is reported that the methanogenic reactor predominantly consist of hydrogen utilizing methanogens and this can be due to greater accessibility of substrates in the methanogenic reactor. The presence of higher amount of fatty acids in methanogenic reactor increase the hydrogen generation which in turn favors the development of hydrogen consumers. Further it is reported that sharing of DNA among the microbes occurs in two-stage digester [41]. The gram negative bacteria possess capabilities to resist antibiotics. The gram negative bacteria in two-stage digester could transfer the antibiotic resistant genes to other microbial population and shift them to be antibiotic resistant microbes.

Pathways for two-stage AD (microbial level)

The phase separation in two-stage digestion (separation of fermentative and methanogenic reactions) may drive considerable changes in metabolic reactions, degradation pathways and in the production of fermentative metabolites and intermediates. Another essential factor is the microbial population that mediates several metabolic pathways. Table 9.2 represents the role of microbes in various metabolic pathways during 2S-AD. Researchers have reported about several metabolic pathways are involved in sugar and protein degradation, VFA degradation and methane production in two-stage digesters. The role of pathways have been investigated in a two-stage CSTR treating cattle manure [53]. It is found that the bacterium, *Thermotogales* and *Sphingobacteriales* acts synergistically with *Clostridiales* and plays a major role in hydrolysis of cellulose during β sugar consumption pathway. In the same study, is also found the syntrophic acetogenesis, lignocellulose degradation and proteoglycan cleavage pathways exist. In syntrophic acetogenesis, the phylum such as *Firmicutes* and *Proteobacteria* degrades VFA. The lignocellulose degradation pathway was reported to be mediated by the bacterial population such as *Pseudomonas, Acinetobacter, Actinobacteria* and *Advenella*, which degrades the recalcitrant lignin compounds. The Proteoglycan cleavage pathway was mediated by Bacteroides, *Cellulophaga and Flavobacterium* which ferments carbohydrates, lipids and proteins.

Upon investigating the performance of two-stage CSTR treating municipal sludge, by a study [45] found that the occurrence of proteolytic, saccharolytic and acidogenic and syntrophic acetate oxidation pathway. In proteolytic, saccharolytic and acidogenic pathway, *Bacteroidia* and *Prevotellaceae* mediates the fermentation of carbohydrates to VFA. In syntrophic acetate oxidation, the families, *Syntrophaceae* and *Syntrophorhabdaceae* undergoes syntrophic association with their hydrogenotrophic

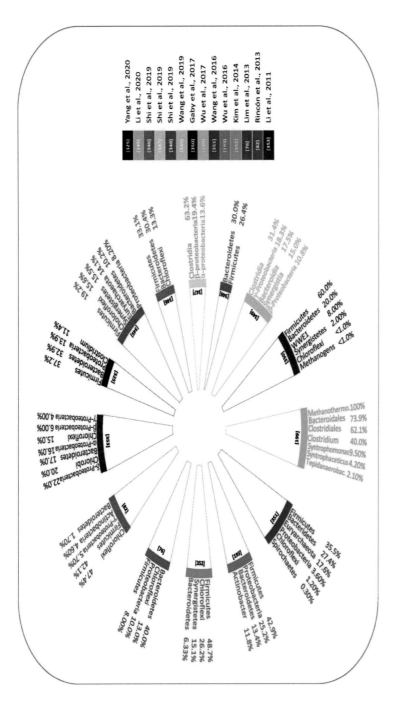

Fig. 9.2 Schematic representation of different microbial communities abundance in two-stage AD

Table 9.1 A summary of microbial communities abundance in various two-stage digesters at varying operational conditions

S. No.	Digester type	Substrate	Operating conditions	Molecular techniques	Comments on microbial community abundance	References
1	Lab scale two-stage digester	Cattle manure	For acidogenic reactor: • Temp—37 °C • HRT—5 days • OLR—6.25 g VS/L day For methanogenic reactor: • Temp—37 °C • HRT—15 days • OLR—6.25 g VS/L day	Ion torrent sequencing	*Firmicutes, Actinobacteria Bacteroidetes* are the predominant bacterial population. Among them, *Firmicutes* are found higher in acidogenic reactor. *Bacteroidetes* and *Actinobacteria* are found higher in methanogenic reactor	[49]
2	Lab scale two-stage digester	Cattle manure	For acidogenic reactor: • Temp—37 °C • HRT—5 days • OLR—6.25 g VS/L day For methanogenic reactor: • Temp—37 °C • HRT—15 days • OLR—6.25 g VS/L day	Illumina sequencing	Prevalence of antibiotic resistant genes in acidogenic reactor	[49]

(continued)

Table 9.1 (continued)

S. No.	Digester type	Substrate	Operating conditions	Molecular techniques	Comments on microbial community abundance	References
3	Lab scale two-stage digester	Cattle manure	For acidogenic reactor: • Temp—37 °C • HRT—5 days • OLR—6.25 g VS/L day For methanogenic reactor: • Temp—37 °C • HRT—15 days • OLR—6.25 g VS/L day	PCR–DGGE analysis	*Acinetobacter*, *Clostridium*, and *Bacillus* are found in abundance. They are highly tolerant to toxic substances	[49]
4	Two-stage CSTR	Food waste and brown water	For acidogenic reactor: • Temp—35 °C • HRT—6 days • OLR—0.5–0.8 g VS/L/d For methanogenic reactor: • Temp—35 °C • HRT—29 days • OLR—0.5–0.8 g VS/L/d	16S rRNA sequencing	The 16S rRNA sequencing revealed that the bacterial community structure of acidogenic reactor were entirely consist of the phyla *Firmicutes* and *Proteobacteria* The bacterial community structure of methanogenic reactor were entirely consist of the phyla *Bacteroidetes*, *Chloroflexi*, *Proteobacteria* and *Firmicutes* in proportions of 40, 13, 10 and 8% of the bacterial clones, respectively	[7]

(continued)

Table 9.1 (continued)

S. No.	Digester type	Substrate	Operating conditions	Molecular techniques	Comments on microbial community abundance	References
5	Two-stage CSTR	Swine manure + Fruits and vegetable waste	For acidogenic reactor: • Temp—55 °C • HRT—3 days For methanogenic reactor: • Temp—55°C • HRT—22 days	DGGE-real time quantitative PCR analysis	Predominance of phylum *Firmicutes* with nine species comes under *clostridiales* and two species comes under *Thermoanaerobacterales* order. *Clostridium cellulosa* thermophilic cellulolytic bacterium predominantly found in acidogenic phase	[14]
6	Two-stage CSTR	Food waste	For acidogenic reactor: • Temp—35 °C • HRT—4 days • OLR—5.7–35.6 g COD/L d For methanogenic reactor: • Temp—35 °C • HRT—12.5 days • OLR-5.7—35.6 g COD/L d	DGGE analysis	This analysis revealed that the acidogenic phase includes bacteria such as *Actinobacteria*, *Firmicutes* (Phyla level), *Aeriscardovia*, *Lactobacillus* (Genera level) The methanogenic phase includes bacteria such as *Firmicutes, Proteobacteria, Spirochaetes and Bacteroidetes* (Phyla); archaea including *Methanomicrobiales, Methanosarcinales and Thermoplasmatales*	[50]
7	Two-stage mesophilic digester	Olive mill residue	For acidogenic reactor: • Temp—35 °C • HRT—12.4 days • OLR—12.9 g COD/L/d For methanogenic reactor: • Temp—35 °C • HRT—4.6 days • OLR—22 g COD/L/d	16S rRNA sequencing and DGGE analysis	Phylum including *Firmicutes, Chloroflexi* (Phyla level); archaea including *Methanosaeta*	[49]

(continued)

Table 9.1 (continued)

S. No.	Digester type	Substrate	Operating conditions	Molecular techniques	Comments on microbial community abundance	References
8	Two-stage hyperthermophilic reactor	Waste activated sludge	• Temp—70 °C • HRT—3 days • OLR—13.62 gCOD/L/d	16S rRNA cloning and sequencing	The hypothermophilic two-stage digester improves phylum percentage such as *Firmicutes* to 27.9%, *Proteobacteria* to 34.4%. The bacteria phyla such as *Bacteroidetes*, *actinobacter* and *Chloroflexi*, were found uniformly distributed in this system The bacterium such as *Methanothermobacter* and *Aceticlastic Methanosaeta* are the predominant genus accounting to 41%	[51]
9	Two-stage mesophilic reactor	Waste activated sludge	• Temp—35 °C • HRT—12 days • OLR—3.32 gCOD/L/d	16S rRNA cloning and sequencing	The mesophilic two-stage digester improves phylum percentage such as methanosarcina to 19%, methanothermobacter to 15.5%. The bacterium such as *Methanosaeta* is the predominant genus accounting to 62.1%	[51]
10	Two-stage CSTR	Cattle manure	For acidogenic reactor: • Temp—35 °C • HRT—25 days For methanogenic reactor: • Temp—35 °C • HRT—33 days	16S rRNA sequencing	*Acidobacteria, Fibrobacteres* and *Planctomycetes* are the oligo member mesophilic bacteria. The predominant species is *Planctomycetia* sp. *DTU247* which has largest genome so they have the capability of digesting significant variety of biopolymers	[52]
11	Two-stage CSTR	Cattle manure	For acidogenic reactor: • Temp—55 °C • HRT—15 days For methanogenic reactor: • Temp—55 °C • HRT—20 days	16S rRNA	*Actinobacteria, Synergistetes* and *Bacteroidetes Thermotogae*, were the dominant population that comes under "oligo-member" phyla and they are inhibited when hydrogen is added externally	[52]

Table 9.2 Role of microbes in various metabolic pathways during two-stage digestion

S. No.	Digester type	Substrate used	Pathway involved	Role of microbes	References
12	Two-stage CSTR	Swine manure and market biowaste	Acetoclastic methanogenesis	Phylum *Firmicutes*, class *Clostridia* utilizes acetate to form methane	[56]
13	Two-stage CSTR	Food waste and brown water	Acetoclastic methanogenesis	*Methanosarcinales* utilizes acetate to form methane	[7]
14	Two-stage CSTR	Food waste and brown water	Starch metabolism	*Lactobacillus amylovorus* showed amylolytic activity. It metabolizes starch to produce lactate and lesser quantity of acetate	[7]
15	Two-stage CSTR	Food waste and brown water	Hydrogenotrophic methanogenesis	*Methanomicrobiales* utilizes hydrogen to form methane	[7]
16	Two-stage mesophilic reactor	Olive mill residue	Acetotrophic methanogenesis	*Methanosaeta* oxidises acetate	[13]
17	Two-stage hyperthermophilic reactor	Waste activated sludge	Protein degradation pathway	*Firmicutes* and *proteobacteria* degrades protein	[24]
18	Two-stage mesophilic reactor	Waste activated sludge	Hydrogenotrophic methanogenesis	*Methanosarcina* and *Methanoculleus* utilizes hydrogen to form methane	[24]
19	Two-stage CSTR	Municipal sludge	Proteolytic, saccharolytic and acidogenic pathway	*Bacteroidia* family, *Prevotellaceae* involved in hydrolysis of carbohydrates and VFA accumulation	[28]
20	Two-stage CSTR	Municipal sludge	Syntrophic acetate oxidation	*Syntrophaceae* and *Syntrophorhabdaceae* form syntrophic association with their hydrogenotrophic partners and oxidizes short chain fatty acids	[28]

(continued)

Table 9.2 (continued)

S. No.	Digester type	Substrate used	Pathway involved	Role of microbes	References
21	Two-stage anaerobic digester	Food waste recycling wastewater	Propionate metabolism	*Desulfotomaculum*, bacteria oxidizes propionate	[33]
22	Two-stage anaerobic digester	Food waste recycling wastewater	Hydrogenotrophic and aceticlastic methanogenesis	*Methanosarcina* are responsible for this pathway	[33]
23	Two-stage anaerobic digester	Food waste recycling wastewater	Syntrophic fatty acid mineralization	*Methanoculleus* degrades acetogenic products	[33]
24	Two-stage CSTR	Cattle manure	β sugar consumption pathway	*Thermotogales* and *Sphingobacteriales* acts synergistically with *Clostridiales* and its role is essential for hydrolysis of cellulose	[34]
25	Two-stage CSTR	Cattle manure	Syntrophic acetogenesis	*Firmicutes* and *Proteobacteria* degrades VFA	[34]
26	Two-stage CSTR	Cattle manure	Lignocellulose degradation pathway	*Pseudomonas*, *Acinetobacter* phylum *Actinobacteria* and genus *Advenella* degrades recalcitrant compounds	[34]
27	Two-stage CSTR	Cattle manure	Proteoglycan cleavage pathway	*Bacteroides*, *Cellulophaga* and *Flavobacterium* ferments carbohydrates, lipids and proteins	[34]
28	Two-stage high solids anaerobic digester	Raw dewatered waste activated sludge	Syntrophic acetogenesis	*Syntrophomonas palmitatica* utilizes straight-chain saturated fatty acids to produce acetate and/or propionate in syntrophic association with the hydrogenotrophic methanogen *Methanospirillum hungatei*	[35]

(continued)

Table 9.2 (continued)

S. No.	Digester type	Substrate used	Pathway involved	Role of microbes	References
29	Two-stage CSTR	Cheese whey permeate and cheese waste powder	Galactose degradation and lactose hydrolysis pathway	The bacterial species *Bifidobacterium crudilactis* degrades the galactose and lactose present on the substrate	[36]
30	Two-stage CSTR	Cheese whey permeate and cheese waste powder	β-oxidation pathway	*Pseudomonas lundensis* produces acetyl CoA and convert into butanoyl-CoA	[36]
31	Two-stage CSTR	Cheese whey permeate and cheese waste powder	β-oxidation pathway	*Syntrophomonas* sp. *UC0014* and *Syntrophaceticus* sp. *UC0017* together with *Pseudomonas lundensis* undergoes syntrophic oxidation of butyrate	[36]
32	Two-stage CSTR	Cheese whey permeate and cheese waste powder	Wood-Ljungdahl (W-L) pathway	*Syntrophaceticus. schinkii* oxidize acetate to hydrogen and/or formate	[36]
33	Two-stage CSTR	Cheese whey permeate and cheese waste powder	Methylotrophic methanogenesis	*Methanomassilicoccus species* produces methane by reducing methanol with hydrogen and by utilizing methylamines	[36]

partners and mediates the oxidation of fatty acids. The presence of acetoclastic methanogenesis pathway in two-stage CSTR codigesting swine market waste where *Firmicutes* and *Clostridia* utilizes acetate to form methane [54]. Syntrophic acetogenesis in two-stage digester have been reported by various studies [53]. However, the bacteria responsible for metabolic reaction varied significantly. In process [55], it was found that the bacterium, *Syntrophomonas palmitatica* utilizes straight-chain saturated fatty acids to generate acetate and/or propionate in syntrophic association with the hydrogenotrophic *methanogen Methanospirillum hungatei*. Ample pathways in two-stage digester treating Cheese whey permeate and cheese waste powder reported [55]. It includes galactose degradation and lactose hydrolysis, β-oxidation, Wood-Ljungdahl (W-L) and Methylotrophic methanogenesis pathway. In galactose degradation and lactose hydrolysis pathway, the bacteria *Bifidobacterium crudilactis* degrades the galactose and lactose content of the feedstock. In β-oxidation pathway, *Syntrophomonas* sp. *UC0014* and *Syntrophaceticus* sp. *UC0017 together* with *Pseudomonas lundensis* undergoes syntrophic oxidation of butyrate. Wood-Ljungdahl (W-L) pathway is another acetate utilizing metabolic pathway where the bacterium, *Syntrophaceticus. schinkii* oxidize acetate to hydrogen and formate. The prevalence of hydrogenotrophic methanogenic pathway was found in methanogenic reactor [39]. During this pathway, *Methanosarcina* and *Methanoculleus* utilizes hydrogen to form methane. Upon investigating pathways in two-stage CSTR treating food waste and brown water, it is reported the occurrence of starch metabolism and acetoclastic methanogenesis pathway [48]. The bacterium, *Lactobacillus amylovorus* showed amylolytic activity. It metabolizes starch to produce lactate and lesser quantity of acetate. In acetoclastic methanogenesis pathway, the bacterium, *Methanosarcinales* utilizes acetate to form methane. Thus in various literatures, it has been reported that the prevalence of metabolic reactions and pathways in various reactors may vary depending upon the substrate and microbes involved.

9.2.2 Process Parameters

A string of operational parameters including low pH, substrate load and shorter hydraulic retention time (HRT) acts as a metabolic buffer that favours the acidogenic phase and are precursors for methanogens in the second stage [55] and is listed in Table 9.3.

Substrate loading

OLR (organic loading rate or substrate loading) typify the quantity of substrate added to the reactor incessantly per day whereas HRT is the period needed to complete the biodegradation process of substrates. OLR beyond the threshold level causes an amassment of volatile fatty acids (VFAs) sequentially depreciating the HRT while protracted retention time leads to inadequate use of organics within the digester [73]. Hence to counterbalance the digestion efficiency and the reactor volume parity

Table 9.3 Performance of various two-stage digesters and its operating conditions

S. No.	Digesters type	Substrate used	Temp (°C)	OLR	HRT (days)	VS removal (%)	Hydrogen yield (or productivity)	Methane yield (or productivity)	Methane (%)	References
34	CSTR + CSTR	FW	38	16	4 + 12	6–15 + 2–5 g VS/L/d	64.1–88.5	371–419 L CH$_4$/kg VS	66.7–74.3	[57]
35	HR + MR	OFMSW	55	39 + 4.16 g COD/L/d	1.9 + 15.4	88.1	11.1 L H$_2$/L$_{fed}$ d	47.4 L CH$_4$/L$_{fed}$ d	NR	[58]
36	CSTR + UASB	OFMSW + FPW	55	8 g VS/L/d + 12 g VS/L/d	3	92.5 + 86.5	NR	NR	NR	[59]
37	UASS + AF	Corn stalk	39	4.5 g VS/L/d	NR	NR	NR	610 mL/g VS	65	[60]
38	Two-stage lab scale anaerobic digestion	Pelletized grass	35	NR	12	NR	6.7 L H$_2$/kg VS	349.4 L CH$_4$/kg VS	NR	[61]
39	Two phase wet anaerobic digestion process	Pre-treated MSW	39	33.0 kg VS/m^3 d	8	78	NR	0.24 N m^3/kg VS	NR	[62]
40	CSTR + CSTR	ES + CW + LCM	37	171.60 + 5.36 kg COD/m^3 d	0.5 + 16	70.02	2.14 L H$_2$/L d	0.90 L CH$_4$/L d	NR	[7]
41	CSTR + CSTR	OMW + CW + LCM	37	126.67 + 3.37 kg COD/m^3 d	0.75 + 25	25.62 + 44.74	1.72 L H$_2$/L$_{reactor}$ d	0.33 L CH$_4$/L$_{reactor}$ d	65.43	[37]
42	Semi-continuous anaerobic sequencing batch reactor	TWW + TSW	38	280 mg COD/L/d	20	NR	NR	251 mL/d	60.5	[39]

(continued)

Table 9.3 (continued)

S. No.	Digesters type	Substrate used	Temp (°C)	OLR	HRT (days)	VS removal (%)	Hydrogen yield (or) productivity	Methane yield (or) productivity	Methane (%)	References
43	CSTR + CSTR	OFMSW	55	NR	1.9	NR	1.077 L H_2/$L_{reactor}$ d	NR	NR	[44]
44	TPAR + TPMR	FVW	35	7.0 kg VS/m^3 d	NR	97.5	NR	0.3 m^3 CH_4/kg VS	NR	[46]
45	AR + MR	VMW	35	4.5 g VS/L/d	15	NR		0.598 LCH_4/g VS_{added}	NR	[47]
46	Mesophilic-Thermophilic Two-Stage	FW	36 + 55	NR	NR	81.7	NR	0.44 L CH_4/g VS_{added}	70.7	[63]
47	CSTR + CSTR	Chicken manure	37 + 53	2.2 g VS/Ld	12	60–67	NR	426–554 mL CH_4/g VS	74	[64]
48	Two-stage lab scale anaerobic digestion	Vinasse	37	NR	NR	64.5	14.8 mL/g $VS_{substrate}$	274 mL/g $VS_{substrate}$	NR	[65]
49	UASB + UASB	Cassava waste water	55	12 kg cod/m^3 d	NR	NR	54.22 mL H_2/$gCOD_{applied}$	356.31 mL CH_4/g MLVSS d	NR	[66]
50	HR + MR	Molasses	35	NR	6 h + 4 d	NR	2.8 L H_2/$L_{reactor}$/d	1.48 L CH_4/$L_{reactor}$/d	NR	[67]
51	LBR + UASB	corn silage	37	5.0 g COD/L/d	4.3	84.4	55.6 ± 6.7 mL/g VS	295.4 ± 14.5 mL/g VS	NR	[68]
52	CSTR + CSTR	FVW + FW	35	4.0 g COD/L/d	10	NR	0.005 L/g VS/d	0.351 L/g VS/d	NR	[69]
53	HR + MR	fw	37	NR	NR	NR	55 mL/g VS	94 mL/g VS	NR	[70]
54	TPAD (Thermophilic–mesophilic)	OFMSW	55 + 35	NR	NR	73.05	NR	2.17 LCH_4/$L_{reactor}$day	NR	[71]

(continued)

Table 9.3 (continued)

S. No.	Digesters type	Substrate used	Temp (°C)	OLR	HRT (days)	VS removal (%)	Hydrogen yield (or) productivity	Methane yield (or) productivity	Methane (%)	References
55	TPAD (Thermophilic–mesophilic)	WAS	65 + 55	15 + 2.2 kg VS/m³ d	2 + 18	55	NR	0.49 m³/kg VS	NR	[72]

EGSB—Expanded Granular Sludge Bed Reactor; CSTR—continuously stirred tank reactors; ES + CW + LCM—ensiled sorghum, + cheese whey + liquid cow manure; OMW + CW + LCM—olive mill wastewater, cheese whey and liquid cow manure; TPAR + TPMR—two-phase acidification reactor + two-phase methanogenic reactor; FVW-Fruit and vegetable waste; AR + MR—acidogenic reactor + methanogenic reactor; FW—food waste; VMW—vegetable market wastes; SuOC—sunflower oil cake SuOC; OFMSW + FPW—organic fraction of the municipal solid waste + food processing waste; TWW + TSW—tannery waste water + tannery solid waste; HR + MR—hydrolytic reactor + methanogenic reactor; TPAD—temperature-phased anaerobic digestion; WAS—waste activated sludge; UASS + AF—up flow anaerobic solid-state reactor + anaerobic filter; LBR + UASB—Leach Bed Reactor + upflow anaerobic sludge blanket; FVW + FW—Fruit and vegetable waste + food waste between the OLR and HRT is required [74, 75]. Thus a two-stage system is found to be propitious ensuing rapid digestion, stable operation and a higher loading capacity.

Optimization of operational parameters such as OLR and HRT is crucial for improved process stability of 2S-AD. Stable condition of a system was evaluated on the basis of pH, VFA, alkalinity whereas, the working condition of a system was reckoned based on everyday gas generation, methane yield, and biodegradation of substrate. An experimental study on a 2S-semi-continuous AD sequential bioreactor was conducted to optimize the working conditions of the methanogenic reactor at equivalent OLR. The findings revealed that COD removal efficiency of 75%, daily biogas of 415 mL/day, methane production of 215 mL/day and methane content about 60.5% was obtained while working under a HRT of 20 days. Also, improved stable performance was witnessed with no VFA accumulation within the optimum operating range. The impact of HRT was investigated [76] on digestion at constant OLRs in a 2S-AD system at mesophilic temperature (35 °C) through pilot scale experiments. It is pointed out that a higher biodegradation percentage and methane generation rate of 33–42% and 22–32% was obtained when retention time was increased with a fixed OLR.

Also, HRT has been laid as a vital parameter influencing microbes environment in two-stage reactors and it has to be optimized emanating a pivotal amount of methane and/or hydrogen [77]. The impact of retention time on H_2 and CH_4 generation through co-digesting olive-mill waste water (OMW), cheese permeate and liquor cattle manure at a mixture proportion of 55:40:5 was assessed using a two-stage CSTR operated under controlled pH at different HRTs in both reactors. The highest hydrogen generation was attained while the acidogenic digester was run at a retention time of 0.75d, whereas highest methane yield was espied at HRT of 25d in the methanogenic reactor. Various studies have revealed that a curtailment in HRT actuates the hydrogen yield as to the lower production rate at higher HRT. This antipode association among the retention time and H_2 generation in acidic reactor was corroborated by Aguilar et al. [78]. In view of the above statement, the existence of homo-acetogens has been foreseen for the limited hydrogen production rate at longer HRTs [63]. The aforementioned fact was further substantiated by a study [79] besides conducting tests in the similar process design and probed the influence of retention time applying the similar blend containing pretreated ensiled sorghum (ES) instead of OMW. The study asserted higher H_2 generation at HRT of 0.5 d in acidogenesis while in methanogenesis at HRT of 16 d highest CH_4 yield was obtained. Conjointly two-stage systems possess the benefit such as buffering the substrate loading stage 1, letting a fixed substrate loading rate in stage 2 [64]. A 2S-AD food waste treating reactor was operated under varying the OLRs at a fixed HRT of 4 days (acidogenic reactor) and 12 days (methanogen reactor). Its performance parameters were and contrasted with a single stage digester [80]. The 2S-AD reactor heightened methane production percentage of 23% and augmented amount of CH_4 from 14 to 71% than the 1S-AD. The impact of OLR and HRT in a 2S-AD bioreactor by conducting the experiments at varying HRT and OLR. Higher methane yield of 0.721 ± 0.010 L/g VS_{add} was procured at a substrate loading of 4.5 g VS/L/d at HRT of 2 and 25 days in both the stage 1 and stage 2 bioreactors. Moreover, achieving elevated substrate loading and lesser retention time in 2S-AD system decreases the digester volume which in turn decreases the capital cost in comparison to single stage reactor. Analogous to the above statement, the influence of the substrate loading and retention time using the biodegradable portion of organic waste and the food processing waste (FPW) was correlated with the traditional 1S and 2S-AD by Aslanzadeh et al. [74]. A healthier reactor performance was achieved at greater OLR and a lesser HRT in the 2S-AD which was assessed at an OLR of 8 g VS/L/d for the FPW and up to 12 g VS/L/d for the biodegradable organic waste, working under a retention time of 3 days. The outcome revealed about 26 and 65% of decreased digester volume in 2S-AD than 1S-AD reactor. The effects of HRT and substrate loading on the hydrolysis and acidogenesis phases of a 2S-AD system was studied [61]. The study outcome divulged that an assortment of OLR affects the organic matter liquefaction slightly; a greater percentage of 30.1% was achieved for a 10 days retention time and a substrate loading of 6 g VS/L/d. Besides, the acidification degree (83.8%) was mainly influenced by the OLR rather than the HRT.

Phase separation is found to be constructive due to the buffering effect of substrate loads in the stage 1 and a consistent condition for methanation process in stage 2. Correspondingly, the above said aspects propagandize the potential of the digester based on substrate biodegradation and methane generation [81]. It is observer in a study [82] that the explored the microbial community studies, substrate biodegradation, and organic acids in thermophilic 2S-AD system at three different OLRs of 5 to 12 kg COD m^{-3} d^{-1} and the results unraveled that the microbial action of hydrolysis and methanogenesis phases enhanced with increment in substrate load while organic matter reduction elevated when the substrate loading increased beyond 5 kg COD m^{-3} d^{-1} and decline at 12 kg COD m^{-3} d^{-1}. Likewise, organic acid reduction of the reactor extricated a declining trend from 96.3, 94.5 and 82% in 1st, 2nd and 3rd substrate loads, respectively. This was due to the insufficient contact time between acetoclastic methanogens and VFA. Two-stage digestion further concludes that treated wastewater recycling from methanogenesis phase to acidogenesis phase can aid the buffering effect of fastly generated organic acids during acetogenesis and uphold an appropriate pH. The treated wastewater from stage 2 bioreactors frequently have acclimated microbes and recirculating them into stage 1 bioreactors amplifies the hydrolysis [4]. A laboratory scale study [4] conducted on a two-stage anaerobic reactor at three different OLRs reviewed the effects of substrate loading and recirculated wastewater on performances of acidic and methanation phases. Under the increasing OLR a substantial buildup of organic acids happened in stage 1 reactor and caused inhibition of hydrolysis. In a correlative study [83] the impact of recycling in organic load increment and retention time decrement was tested in a 2S-AD bioreactor using carbohydrate-based starch and cotton as substrate. Furthermore, the study affirmed that at higher OLR, recirculation struck hydrolysis step averting nutrient loss and thus improvised the process stability and performance.

Influence of temperature

The effectiveness and constancy of digestion system is determined by equivalence among hydrolytic, acidogenic, acetogenic and methanogenic stages. Temperature is one of the censorious parameter that impacts stable performance and microbiological action in determining the operating conditions of the process during AD [84]. Methanogens are normally highly subtle to disparity in temperature than the other microbes that mediates AD [85]. The impact of temperature on biomethanation phase of 2S-AD for converting oil industry wastewater to generate methane was unveiled in a study by Trisakthi et al. [86]. In this study the authors vary the mesophilic temperature (30–42 °C) and thermophilic temperature (43–55 °C) at retention time of 4 days. It was culminated that degradation of organic content, biogas production and its methane content augmented by increase in temperature. At mesophilic temperature, removal of volatile solids and organics was reported to be 51.56 ± 8.30 and 79.82 ± 6.03%. Concurrently at thermophilic range they were 67.44 ± 3.59 and 79.16 ± 1.75%, respectively. Thus thermophilic digestion (50–55 °C) in comparison with mesophilic digestion (37–40 °C) is inclined as a viable option for removing solids and destroying the pathogenic microbes for treating biowaste [65]. But extra energy necessities to uphold elevated temperatures becomes a hitch in thermophilic reactors [66]. Temperature-phased anaerobic digestion (TPAD) system integrates a thermophilic phase having lesser HRT and mesophilic phase showing lengthier HRT and has been recognized to afford destruction of pathogenic microbes, efficient removal of organics and this depends on 2S-AD [87]. A study to determine the efficiency of a semi-continuous TPAD, treating biodegradable organic waste with a single stage system. The TPAD systems exhibited higher removal efficiencies of organic matter (16, 10 and 30% for total organic carbon, soluble organics and volatile solids) as well as greater biogas productivity (26–60%) for 4 days retention time in thermophilic digester and 10 days retention time in mesophilic digester than those of the single reactor systems, for similar retention times. A study [67] evaluated the two phase configurations: mesophilic-mesophilic, thermophilic–mesophilic and mesophilic–thermophilic on the basis of solids reduction, reduction of organics, and methane generation. It was conceded that the mesophilic-thermophilic outperformed the other systems. A study [88] investigated the generation of methane using poultry waste as substrate at various substrate loads (1.9–4.7g VS/L/d), in a mesophilic-thermophilic 2S-AD. The reactors with mesophilic temperature showed increased substrate biodegradation in comparison to thermophilic reactors. In another feasibility study of 2S-AD (UASS and AF) treating corn stalk, the process performance parameters were compared to 1S and 2S—system under mesophilic and thermophilic temperatures at different OLRs (2.5, 4.5, and 8.0 g VS L/1 d/1). The mesophilic 2S-AD systems showed greater biogas generation than the 1S-AD and thermophilic 2S-AD systems. This was because of the better degradation of lignocellulosic biomass at elevated temperature [89].

Substrates, productivity and yield

It is impossible to compare directly the methane generation from various substrates. This is because the production result for particular substrates are obtained in a varied operational circumstances. Thereupon, substrates are placed in association with eventual biomethane production evaluated through biomethanation tests [90]. A variety of organic materials such as plant wastes, manure of animals, effluents with greater organics, agro and agricultural industries residuals have been studied and documented for methane production [18]. The Carbon/Nitrogen proportion echoes the concentration and quantity of substrate, and therefore, reactors are subtle to Carbon/Nitrogen proportion [73].

Substrates with less Carbon/Nitrogen proportion upsurge the issue of ammonia accumulation, and this is lethal to methanogenic microbes. This in turn leads to inadequate application of nutrients. Sugars were observed to lessen the issues connected with inadequate substrate sources [68]. The instantaneous generation of biohydrogen and biomethane typically utilizes sugar rich substrates, cane molasses and sugar syrup [91, 92], cassava effluent [93] and bagasses [94]. Biogasification performance of vinasse was probed through two-stage anaerobic digestion which revealed 10.8% higher methane yield than that of one-stage [95]. A study [96] investigated H_2 and CH_4 generation from cassava effluent via 2S-AD bioreactor, UASB and retrieved the greatest H_2 and CH_4 yields of 80.25 mL H_2/g COD and 183.31 mL CH_4/gCOD was found at the same optimum COD loading rate. Consistent outcomes were procured in a study [69] at same conditions and reaped maximal H_2 and CH_4 yields of 15 mL H_2/g COD and 259 mL CH_4/g COD at an optimal substrate load of 1200 mg/L. A study [97] used molasses as substrates for the 2S-AD and acquired highest H_2 and CH_4 yields of 2.8 L-H_2/L-reactor/d and 1.48 L-CH_4/L-reactor/d. Crop residues typically contain a high lignocellulosic content [98]. Applying agro wastes as such or combining with any other wastes leads to lesser methane production owing to elevated proportion of Carbon/Nitrogen and inhibitory compound, lignin [99]. 2S-AD (leach bed-UASB reactor) digesting corn fodder and cattail leads to the redistribution of biomethane generation in the 2S process showing 3–10% increment in the leach bed digester and coaxed a reduction in digestion time [100]. The existence of lignin and cellulose containing plants employing mixed consortia in a 2S-AD was explored for their H_2 and CH_4 yields revealing higher production rates and yield than single stage digester [101, 102]. A study [103] run by a 2S-CSTR obtained a CH_4 production of 419 NL CH_4/kg VS by utilizing food waste as substrate and is about 23% greater when compared to 1S-AD. Biodegradation of proteinaceous compounds causes discharge of NH_3 ions, that are heavily inhibitory to methanogens [79]. However, FW which contains high protein and lipid content causes issues associated with mass transfer and inhibition of microbes [104]. Appropriate co-substrates and optimal Carbon/Nitrogen proportion or bioaugmentation of microbes in bioreactor could diminish this issue [105]. In a similar study, co-digestion of market and food waste proffered higher CH_4 production of 7.0–15.8% in a 2S-AD at higher level of OLRs (>2.0 g (VS) L^{-1} d^{-1}) when compared to 1S-AD [106].

Effect of Mixing

Anaerobic digesters stability is greatly influenced by: (1) the extent of interaction among the microbes and the feedstocks, and (2) the contact among methanogenic and their syntrophic microbes and on the whole it is the role of mixing pattern in digester. However, mixing effect on community structure of microbes and reactor parameters are ambiguous yet [89]. Consequently, the adequate mixing was backed in numerous reports [82, 107, 108] and, concurrently, put to question by many others [109, 110].

The mutation in operational temperature of reactor impinges the microbial action of the reactor especially the methnaogens. Suffice mixing of the substrate and inoculum averts the alteration of temperature [111]. The recirculating effect digested residue is an essential operation factor affecting hydrogen and methane generation in 2S-AD system fed on organic solids. In addition, it is suggested that the recirculating effect of effective methanogenic effluent and this showed inhibitory impact on H_2 yield in stage 1 digester [112]. The recirculating effect can maintain organic acids accumulation, alkalinity, and methanogenic microbes in equilibrium in stage 1 and stage 2 reactors and hence enhance the stability of bioreactors [113]. The rate of recirculation have impact on the stability of 2S-AD [114]. The results pinpointed the dynamics among hydrolytic and methanogenic phases via recirculating effect and specified efficient mass transfer potential among 2S-AD reactors. [115].

9.3 Mass and Heat Transfer

The 2S-AD process was typically propound for fortifying stable working, good system design which resorts to digesters for acetogenesis and biomethanation phases, in agreement with optimizing both the phases [113, 115, 116]. The concentration and composition of VFAs profoundly relies on the feed characteristics and working parameters [117]. Mixing allows for close interaction among microbes and substrates, and exalts mass transfers of intermediary compounds inside the reactors for effectual digestion. Quite a few investigators have labored CSTR for stage 1 reactors to pare this entity and heighten the biodegradation of substrates [118]. Entire mixing could cause disruption in separating liquid from solids, a phase essential in effectual feeding of liquor and liquefied organic matter into the stage 2 reactor. In a first phase reactor, albeit a revolving cylinder net sieve has been fixed for mixing and separating liquor and solids synchronously [119], lofty input of energy and convoluted production persist as a worry to process charges. The effluent recirculation from the stage 2 to the stage 1 bioreactor is the menacing processes and regulatory techniques that employs ample clout on the entire system stability. Recirculation makes an enticing gas yield in which the pH attained an appropriate value because of the buffering effect of the recirculating effluent [118]. In a study it was found that optimized recirculation rate could revamp hydrolytic phase in stage 1 digester. The equilibrium of organic acids in the entire 2S-AD process are disturbed through virtue of elevated recirculation rate further the elevated recirculating extent catered to faster mixing of liquor among 2S-AD system and these rapid circulation epitomizes a diminution in mass transfer limitation in aqueous phase leading to a slump in the system efficiency [120]. Diversely, high recirculation rate does not afford sufficient period for biodegradation of organic acids through the microbes and henceforth culminated to its accretion in the methanogenic reactor which precipitates a disproportionate inflation in the efficient substrate load of the stage 2 (methane producing) bioreactor waning its performance. The biological pathways starts from mass transfer of substrate to microbes in various bioreactor designs during varying operational circumstances

[121]. A complete modeling of a 2S-AD was established by Yu et al. [122] to study the process interface among a high solid anaerobic digester (HSAD) and UASB reactor in a pilot-scale study. The organic solid waste were crushed in mechanical mode to amplify mass transfer between microbes, enzymatic activity and solid feed. The estimates illustrated that recirculated methanogenesis microbes elevated yield of methane and ebbed yield of hydrogen in stage 1 digester at increased recirculation extent. A substrate load of 5 kg COD m^{-3} day^{-1} sustained incessant methane yield and it can be improved through the amplified UASB unit area and recirculation of effluent from methanogenesis bioreactor. The input of the amplified UASB unit area for dropping accumulation of organic acids could be ascribed to lesser speed, i.e. lesser mixing power. It can be confirmed from the literature that dynamic mixing is incongruous in a bioreactor in which methane producing phase is the rate restricting phase, as only deficient mixing provides the spots where methanogens can be sheltered from fast acidic phase [123]. Berhe et al. [124] evaluated the process stability of the methanogenic step in a 2S-ASBR codigesting tannery solid and liquid waste. Here, when the ratio of TSW exceeds 50% a slackening of daily biogas production was detected owing to the accumulation of solids in the digesters, deterring the mixing of acidogenic bioreactor contents and causing difficulty in mass transfer efficiency. A subsequent attrition in the VFA production emanating less methane generation in the particular methane producing bioreactor was contemplated. This specifies that the substrate concentration regulates the efficiency of microbial biomass [125]. In addition, it is explored that the thermophilic 2S-AD of organic waste for hydrogen and methane generation at varying substrates loading and the outcomes concerned the role of recirculating effect on pH alteration [126]. The results displays that mass transfer in greater solid content can be augmented at thermophilic temperature. The viability of employing 2S system with structured-bed reactor, precisely as stage 2 methane producing bioreactor for treating sugarcane industry waste was demonstrated [127]. The domination of microbes in immobilization stage along with the acclimated sludge in thermophilic temperature resulted in increased substrate biodegradation. On the other hand, by employing the same environment in UASB digester leads to system failure, owing to mass transfer limitation resulting from elevated concentration of inert solids.

9.3.1   Advantages and Limitations

Customarily legion researchers have consented with the fact that 2S-AD bioreactors working in 2S design are callable to 1S-AD based on biogas generation and stable biodegradation. The two-stage configuration amplifies stable performance in terms of diverse microbes against fluctuations [6]. Besides, the first acidic bioreactor can serve as an efficient buffer contrary to hasty pH decline engendering VFAs aggregation and hampers methanogenic microorganisms [5]. Appropriately, greater system dependability, flexibility, stable performance and greater biodegradation are foreknown in 2S-AD. In addition, these reactors can be put to use to biodegrade effluent with greater organics and stalwart environments such as low temperature [128]. 2S-AD possess the asset of improved bioenergy generation, greater organic biodegradation and biogas production as to single-stage [129] and a higher organic loading rate (OLR) [130]. A 2S-AD system working under both meso and thermophilic range of temperature treating sludge, accrued greater reduction of organics and solids when compared to 1S-AD [131]. Thermophilic-Mesophilic anaerobic digesters undeniably leverages increased percentage of pathogen removal, higher reduction of biosolids, and elevated methane generation, which is ascertained in wastewater treatment [132]. The 2S-AD is applicable for treating typical biowastes that possess greater amount of lipids [133]. Besides, a study investigated the 2S-expanded granule bed bioreactor cogently reduced organic matter and sulphur compounds [134]. A two-year statistical analysis [7] on 2S-AD digester, reported that the digester can sustain stable performance at lesser substrate loading (0.79 $\pm$ 0.16 kg-COD/m^3·d). The predominant microbes during the stage 1 and 2 are *Clostridium sporogenes* and *Methanobacteriales* in 2S-AD [135]. The above said characteristic enable to circumvent the effect of excess load of substrate in stage 2 reactor at biomethanation phase. Moreover, H2 can be generated utilizing organic matter, organic acids and alcohol in stage 1 and it leads to stabilization of organic waste. However, a few shortcomings of 2S-AD are divulged which includes (i) accumulation of H2 which inhibits acidogens; (ii) exclusion of conceivable reliant substrate need for methanogens; (iii) practical difficulty and (iv) elevated investment for starting up [136]. The 2S-AD disentangles the problem in conjunction with organic acids accumulation, and also aids to investigate the microbes and contradistinctions in the digester, regardless of elevated operation cost and convoluted amenities [99].

References

1. Stabnikova O, Liu XY, Wang JY (2008) Digestion of frozen/thawed food waste in the hybrid anaerobic solid–liquid system. Waste Manag 28:1654–1659
2. Kumar A, Yadav AK, Sreekrishnan TR, Satya S, Kaushik CP (2008) Treatment of low strength industrial cluster wastewater by anaerobic hybrid reactor. Bioresour Technol 99:3123–3129
3. Kundu K, Sharma S, Sreekrishnan TR (2012) Effect of operating temperatures on the microbial community profiles in a high cell density hybrid anaerobic bioreactor. Bioresour Technol 118:502–511
4. Venetsaneas N, Antonopoulou G, Stamatelatou K, Kornaros M, Lyberatos G (2009) Using cheese whey for hydrogen and methane generation in a two-stage continuous process with alternative pH controlling approaches. Bioresour Technol 100:3713–3717
5. Xie L, Dong N, Wang L, Zhou Q (2014) Thermophilic hydrogen production from starch wastewater using two-phase sequencing batch fermentation coupled with UASB methanogenic effluent recycling. Int J Hydrog Energy 39:20942–20949
6. Akyol Ç, Aydin S, Ince O, Ince B (2016) A comprehensive microbial insight into single-stage and two-stage anaerobic digestion of oxytetracycline-medicated cattle manure. Chem Eng J 303:675–684
7. Dareioti MA, Kornaros M (2015) Anaerobic mesophilic co-digestion of ensiled sorghum, cheese whey and liquid cow manure in a two-stage CSTR system: effect of hydraulic retention time. Bioresour Technol 175:553–562

8. Hafez H, Nakhla G, El Naggar H (2010) An integrated system for hydrogen and methane production during landfill leachate treatment. Int J Hydrog Energy 35:5010–5014

9. Antonopoulou G, Stamatelatou K, Venetsaneas N, Kornaros M, Lyberatos G (2008) Biohydrogen and methane production from cheese whey in a two-stage anaerobic process. Ind Eng Chem Res 47:5227–5233

10. Sabry T (2008) Application of the UASB inoculated with flocculent and granular sludge in treating sewage at different hydraulic shock loads. Bioresour Technol 99:4073–4077

11. Cota-Navarro C, Carrillo-Reyes J, Davila-Vazquez G, Alatriste-Mondragón F, Razo-Flores E (2011) Continuous hydrogen and methane production in a two-stage cheese whey fermentation system. Water Sci Technol 64:367–374

12. O-Thong S, Suksong W, Promnuan K, Thipmunee M, Mamimin C, Prasertsan P (2016) Two-stage thermophilic fermentation and mesophilic methanogenic process for biohythane production from palm oil mill effluent with methanogenic effluent recirculation for pH control. Int J Hydrog Energy 41:21702–21712

13. Kisielewska M, Wysocka I, Rynkiewicz MR (2014) Continuous biohydrogen and biomethane production from whey permeate in a two-stage fermentation process. Environ Prog Sustain Energy 33:1411–1418

14. Kongjan P, O-Thong S, Angelidaki I (2011) Performance and microbial community analysis of two-stage process with extreme thermophilic hydrogen and thermophilic methane production from hydrolysate in UASB reactors. Bioresour Technol 102:4028–4035

15. Mamimin C, Singkhala A, Kongjan P, Suraraksa B, Prasertsan P, Imai T, O-Thong S (2015) Two-stage thermophilic fermentation and mesophilic methanogen process for biohythane production from palm oil mill effluent. Int J Hydrog Energy 40:6319–6328

16. Elreedy A, Tawfik A, Kubota K, Shimada Y, Harada H (2015) Hythane (H2 + CH4) production from petrochemical wastewater containing mono-ethylene glycol via stepped anaerobic baffled reactor. Int Biodeterior Biodegrad 105:252–261

17. Shanmugam AS, Akunna JC (2010) Modelling head losses in granular bed anaerobic baffled reactors at high flows during start-up. Water Res 44:5474–5480

18. Ahamed A, Chen C-L, Rajagopal R, Wu D, Mao Y, Ho IJR, Lim JW, Wang JY (2015) Multi-phased anaerobic baffled reactor treating food waste. Bioresour Technol 182:239–244

19. Álvarez JA, Armstrong E, Gómez M, Soto M (2008) Anaerobic treatment of low-strength municipal wastewater by a two-stage pilot plant under psychrophilic conditions. Bioresour Technol 99:7051–7062

20. Wu L-J, Kobayashi T, Li Y-Y, Xu K-Q (2015) Comparison of single-stage and temperature-phased two-stage anaerobic digestion of oily food waste. Energy Convers Manag 106:1174–1182

21. Li Y, Liu H, Yan F, Su D, Wang Y, Zhou H (2017) High-calorific biogas production from anaerobic digestion of food waste using a two-phase pressurized biofilm (TPPB) system. Bioresour Technol 224:56–62

22. Yun Y, Sung S, Shin H, Han J, Kim H, Kim D (2017) Producing desulfurized biogas through removal of sulfate in the first-stage of a two-stage anaerobic digestion. Biotechnol Bioeng 114:970–979

23. Wang TX, Ma XY, Wang MM, Chu HJ, Zuo JE, Yang YF (2016) A comparative study of microbial community compositions in thermophilic and mesophilic sludge anaerobic digestion systems. Microbiol China 43:26–35

24. Wu L-J, Higashimori A, Qin Y, Hojo T, Kubota K, Li Y-Y (2016) Comparison of hyper-thermophilic–mesophilic two-stage with single-stage mesophilic anaerobic digestion of waste activated sludge: process performance and microbial community analysis. Chem Eng J 290:290–301

25. Shimada T, Morgenroth E, Tandukar M, Pavlostathis SG, Smith A, Raskin L, Kilian RE (2011) Syntrophic acetate oxidation in two-phase (acid–methane) anaerobic digesters. Water Sci Technol 64:1812–1820

26. Li W-W, Yu H-Q (2011) From wastewater to bioenergy and biochemicals via two-stage bioconversion processes: a future paradigm. Biotechnol Adv 29:972–982

27. Maspolim Y, Zhou Y, Guo C, Xiao K, Ng WJ (2015) Determination of the archaeal and bacterial communities in two-phase and single-stage anaerobic systems by 454 pyrosequencing. J Environ Sci 36:121–129

28. Maspolim Y, Zhou Y, Guo C, Xiao K, Ng WJ (2015) Comparison of single-stage and two-phase anaerobic sludge digestion systems—performance and microbial community dynamics. Chemosphere 140:54–62

29. Meng Y, Jost C, Mumme J, Wang K, Linke B (2016) An analysis of single and two stage, mesophilic and thermophilic high-rate systems for anaerobic digestion of corn stalk. Chem Eng J 288:79–86

30. Böske J, Wirth B, Garlipp F, Mumme J, Van den Weghe H (2015) Upflow anaerobic solid-state (UASS) digestion of horse manure: thermophilic vs. mesophilic performance. Bioresour Technol 175:8–16

31. Goswami R, Chattopadhyay P, Shome A, Banerjee SN, Chakraborty AK, Mathew AK, Chaudhury S (2016) An overview of physico-chemical mechanisms of biogas production by microbial communities: a step towards sustainable waste management. 3 Biotech 6:72

32. Ziemiński K, Frąc M (2012) Methane fermentation process as anaerobic digestion of biomass: transformations, stages and microorganisms. Afr J Biotechnol 11:4127–4139

33. Shin SG, Han G, Lim J, Lee C, Hwang S (2010) A comprehensive microbial insight into two-stage anaerobic digestion of food waste-recycling wastewater. Water Res 44:4838–4849

34. Bassani I, Kougias PG, Treu L, Angelidaki I (2015) Biogas upgrading via hydrogenotrophic methanogenesis in two-stage continuous stirred tank reactors at mesophilic and thermophilic conditions. Environ Sci Technol 49:12585–12593

35. Wu J, Cao Z, Hu Y, Wang X, Wang G, Zuo J, Wang K, Qian Y (2017) Microbial insight into a pilot-scale enhanced two-stage high-solid anaerobic digestion system treating waste activated sludge. Int J Environ Res Public Health 14:1483

36. Fontana A, Campanaro S, Treu L, Kougias PG, Cappa F, Morelli L, Angelidaki I (2018) Performance and genome-centric metagenomics of thermophilic single and two-stage anaerobic digesters treating cheese wastes. Water Res 134:181–191

37. Dareioti MA, Kornaros M (2014) Effect of hydraulic retention time (HRT) on the anaerobic co-digestion of agro-industrial wastes in a two-stage CSTR system. Bioresour Technol 167:407–415

38. Aslanzadeh S (2014) Pretreatment of cellulosic waste and high rate biogas production. University of Borås, School of Engineering

39. Berhe S, Leta S (2018) Anaerobic co-digestion of tannery waste water and tannery solid waste using two-stage anaerobic sequencing batch reactor: focus on performances of methanogenic step. J Mater Cycles Waste Manag 2018:1–15

40. Xiao B, Qin Y, Wu J, Chen H, Yu P, Liu J, You Li Y (2018) Comparison of single-stage and two-stage thermophilic anaerobic digestion of food waste: performance, energy balance and reaction process. Energy Convers Manag 156:215–223

41. Merlino G, Rizzi A, Schievano A, Tenca A, Scaglia B, Oberti R, Adani F, Daffonchio D (2013) Microbial community structure and dynamics in two-stage vs single-stage thermophilic anaerobic digestion of mixed swine slurry and market bio-waste. Water Res 47:1983–1995

42. Rincón B, Portillo MC, González JM, Borja R (2013) Microbial community dynamics in the two-stage anaerobic digestion process of two-phase olive mill residue. Int J Environ Sci Technol 10:635–644

43. Ghorbanian M, Lupitskyy RM, Satyavolu JV, Berson RE (2014) Impact of supplemental hydrogen on biogas enhancement and substrate removal efficiency in a two-stage expanded granular sludge bed reactor. Environ Eng Sci 31:253–260

44. Aguilar MAR, Fdez-Güelfo LA, Álvarez-Gallego CJ, García LIR (2013) Effect of HRT on hydrogen production and organic matter solubilization in acidogenic anaerobic digestion of OFMSW. Chem Eng J 219:443–449

45. Chen W-H, Sung S, Chen S-Y (2009) Biological hydrogen production in an anaerobic sequencing batch reactor: pH and cyclic duration effects. Int J Hydrogen Energy 34:227–234

46. Ganesh R, Torrijos M, Sousbie P, Lugardon A, Steyer JP, Delgenes JP (2014) Single-phase and two-phase anaerobic digestion of fruit and vegetable waste: comparison of start-up, reactor stability and process performance. Waste Manag 34:875–885

47. Sridevi VD, Rema T, Srinivasan SV (2015) Studies on biogas production from vegetable market wastes in a two-phase anaerobic reactor. Clean Technol Environ Policy 7:1689–1697

48. Lim JW, Chen C-L, Ho IJR, Wang J-Y (2013) Study of microbial community and biodegradation efficiency for single-and two-phase anaerobic co-digestion of brown water and food waste. Bioresour Technol 147:193–201

49. Aboudi K, Quiroga XG, Gallego CJA, García LIR (2017) Comparison of single-stage and temperature-phased anaerobic digestion of sugar beet by-products. In: 5th international conference on sustainable solid waste management, 24–27 June, 2017, p 1–24

50. Moon HC, Song IS (2011) Enzymatic hydrolysis of foodwaste and methane production using UASB bioreactor. Int J Green Energy 8:361–371

51. Wan S, Sun L, Sun J, Luo W (2013) Biogas production and microbial community change during the Co-digestion of food waste with Chinese silver grass in a single-stage anaerobic reactor. Biotechnol Bioprocess Eng 18:1022–1030

52. Treu L, Campanaro S, Kougias PG, Sartori C, Bassani I, Angelidaki I (2018) Hydrogen-fueled microbial pathways in biogas upgrading systems revealed by genome-centric metagenomics. Front Microbiol 9:1079

53. Wijekoon KC, Visvanathan C, Abeynayaka A (2011) Effect of organic loading rate on VFA production, organic matter removal and microbial activity of a two-stage thermophilic anaerobic membrane bioreactor. Bioresour Technol 102:5353–5360

54. Schievano A, Tenca A, Scaglia B, Merlino G, Rizzi A, Daffonchio D, Oberti R, Adani F (2012) Two-stage vs single-stage thermophilic anaerobic digestion: comparison of energy production and biodegradation efficiencies. Environ Sci Technol 46:8502–8510

55. Tomei MC, Bertanza G, Canato M, Heimersson S, Laera G, Svanström M (2016) Techno-economic and environmental assessment of upgrading alternatives for sludge stabilization in municipal wastewater treatment plants. J Clean Prod 112:3106–3115

56. He Y, Pang Y, Liu Y, Li X, Wang K (2008) Physicochemical characterization of rice straw pretreated with sodium hydroxide in the solid state for enhancing biogas production. Energy Fuels 22:2775–2781

57. Yu L, Wensel PC, Ma JW, Chen SL (2014) Mathematical modeling in anaerobic digestion (AD). J Bioremediation Biodegrad 5:003

58. Adekunle KF, Okolie JA (2015) A review of biochemical process of anaerobic digestion. Adv Biosci Biotechnol 6:205

59. Karim K, Thoma GJ, Al-Dahhan MH (2007) Gas-lift digester configuration effects on mixing effectiveness. Water Res 41:3051–3060

60. Banu JR, Arulazhagan P, Kumar SA, Kaliappan S, Lakshmi AM (2015) Anaerobic co-digestion of chemical- and ozone-pretreated sludge in hybrid upflow anaerobic sludge blanket reactor. Desalin Water Treat 54:3269–3278

61. Massanet-Nicolau J, Dinsdale R, Guwy A, Shipley G (2015) Utilising biohydrogen to increase methane production, energy yields and process efficiency via two stage anaerobic digestion of grass. Bioresour Technol 189:379–383

62. De Gioannis G, Diaz LF, Muntoni A, Pisanu A (2008) Two-phase anaerobic digestion within a solid waste/wastewater integrated management system. Waste Manag 28:1801–1808

63. Ventura J-RS, Lee J, Jahng D (2014) A comparative study on the alternating mesophilic and thermophilic two-stage anaerobic digestion of food waste. J Environ Sci 26:1274–1283

64. Dalkılıc K, Ugurlu A (2015) Biogas production from chicken manure at different organic loading rates in a mesophilic-thermopilic two stage anaerobic system. J Biosci Bioeng 120:315–322

65. Fu SF, Xu XH, Dai M, Yuan XZ, Guo RB (2017) Hydrogen and methane production from vinasse using two-stage anaerobic digestion. Process Saf Environ Prot 107:81–86

66. Intanoo P, Rangsanvigit P, Malakul P, Chavadej S (2014) Optimization of separate hydrogen and methane production from cassava wastewater using two-stage upflow anaerobic sludge

blanket reactor (UASB) system under thermophilic operation. Bioresour Technol 173:256–265

67. Park MJ, Jo JH, Park D, Lee DS, Park JM (2010) Comprehensive study on a two-stage anaerobic digestion process for the sequential production of hydrogen and methane from cost-effective molasses. Int J Hydrog Energy 35:6194–6202

68. Nkemka VN, Gilroyed B, Yanke J, Gruninger R, Vedres D, McAllister T, Hao X (2015) Bioaugmentation with an anaerobic fungus in a two-stage process for biohydrogen and biogas production using corn silage and cattail. Bioresour Technol 185:79–88

69. Shen F, Yuan H, Pang Y, Chen S, Zhu B, Zou D, Liu Y, Ma J, Yu L, Li X (2013) Performances of anaerobic co-digestion of fruit & vegetable waste (FVW) and food waste (FW): single-phase vs. two-phase. Bioresour Technol 144:80–85

70. Nathao C, Sirisukpoka U, Pisutpaisal N (2013) Production of hydrogen and methane by one and two stage fermentation of food waste. Int J Hydrog Energy 38:15764–15769

71. Fernández-Rodríguez J, Pérez M, Romero LI (2016) Semicontinuous temperature-phased anaerobic digestion (TPAD) of organic fraction of municipal solid waste (OFMSW). Comparison with single-stage processes. Chem Eng J 285:409–416

72. Bolzonella D, Cavinato C, Fatone F, Pavan P, Cecchi F (2012) High rate mesophilic, thermophilic, and temperature phased anaerobic digestion of waste activated sludge: a pilot-scale study. Waste Manag 32:1196–1201

73. Andalib M, Elbeshbishy E, Mustafa N, Hafez H, Nakhla G, Zhu J (2014) Performance of an anaerobic fluidized bed bioreactor (AnFBR) for digestion of primary municipal wastewater treatment biosolids and bioethanol thin stillage. Renew Energy 71:276–285

74. Aslanzadeh S, Rajendran K, Taherzadeh MJ (2014) A comparative study between single- and two-stage anaerobic digestion processes: effects of organic loading rate and hydraulic retention time. Int Biodeterior Biodegrad 95:181–188

75. Trisakti B, Irvan M, Turmuzi M (2017) Effect of temperature on methanogenesis stage of two-stage anaerobic digestion of palm oil mill effluent (POME) into biogas. In: IOP conference series: materials science and engineering, vol 206. IOP Publishing, p 12027

76. Fernández-Rodríguez J, Pérez M, Romero LI (2013) Comparison of mesophilic and thermophilic dry anaerobic digestion of OFMSW: kinetic analysis. Chem Eng J 232:59–64

77. Schnurer A, Jarvis A (2010) Microbiological handbook for biogas plants. Swedish Waste Manag U 2009:1–74

78. Riau V, De la Rubia MÁ, Pérez M (2010) Temperature-phased anaerobic digestion (TPAD) to obtain class A biosolids: a semi-continuous study. Bioresour Technol 101:2706–2712

79. Kafle GK, Kim SH (2011) Sludge exchange process on two serial CSTRs anaerobic digestions: process failure and recovery. Bioresour Technol 102:6815–6822

80. Voelklein MA, Jacob A, O'Shea R, Murphy JD (2016) Assessment of increasing loading rate on two-stage digestion of food waste. Bioresour Technol 202:172–180

81. Zhang T, Liu L, Song Z, Ren G, Feng Y, Han X, Yang G (2013) Biogas production by co-digestion of goat manure with three crop residues. PLoS One 8:e66845

82. Cavinato C, Bolzonella D, Fatone F, Cecchi F, Pavan P (2011) Optimization of two-phase thermophilic anaerobic digestion of biowaste for hydrogen and methane production through reject water recirculation. Bioresour Technol 102:8605–8611

83. Baêta BEL, Lima DRS, Balena Filho JG, Adarme OFH, Gurgel LVA, de Aquino SF (2016) Evaluation of hydrogen and methane production from sugarcane bagasse hemicellulose hydrolysates by two-stage anaerobic digestion process. Bioresour Technol 218:436–446

84. Nualsri C, Kongjan P, Reungsang A (2016) Direct integration of CSTR-UASB reactors for two-stage hydrogen and methane production from sugarcane syrup. Int J Hydrog Energy 41:17884–17895

85. Intanoo P, Chaimongkol P, Chavadej S (2016) Hydrogen and methane production from cassava wastewater using two-stage upflow anaerobic sludge blanket reactors (UASB) with an emphasis on maximum hydrogen production. Int J Hydrog Energy 41:6107–6114

86. Abd-Alla MH, Bagy MMK, Morsy FM, Hassan EA (2014) Enhancement of biodiesel, hydrogen and methane generation from molasses by Cunninghamella echinulata and anaerobic bacteria through sequential three-stage fermentation. Energy 78:543–554

87. Wangmor T (2014) Effect of added cassava residue on hydrogen and methane production from cassava wastewater using a two stage UASB system. In: 5th research symposium on petrochemical and materials technology and the 20th ppc symposium on petroleum, petrochemicals, and polymers. Bangkok, pp 217–325

88. Chen Y, Cheng JJ, Creamer KS (2008) Inhibition of anaerobic digestion process: a review. Bioresour Technol 99:4044–4064

89. Hagos K, Zong J, Li D, Liu C, Lu X (2017) Anaerobic co-digestion process for biogas production: progress, challenges and perspectives. Renew Sustain Energy Rev 76:1485–1496

90. Veeken A, Hamelers B (1999) Effect of temperature on hydrolysis rates of selected biowaste components. Bioresour Technol 69:249–254

91. Ko J-J, Shimizu Y, Ikeda K, Kim S-K, Park C-H, Matsui S (2009) Biodegradation of high molecular weight lignin under sulfate reducing conditions: lignin degradability and degradation by-products. Bioresour Technol 100:1622–1627

92. Akobi C, Yeo H, Hafez H, Nakhla G (2016) Single-stage and two-stage anaerobic digestion of extruded lignocellulosic biomass. Appl Energy 184:548–559

93. Jeihanipour A, Aslanzadeh S, Rajendran K, Balasubramanian G, Taherzadeh M (2013) High-rate biogas production from waste textiles using a two-stage process. Renew Energy 52:128–135

94. Schnürer A, Nordberg Å (2008) Ammonia, a selective agent for methane production by syntrophic acetate oxidation at mesophilic temperature. Water Sci Technol 57:735–740

95. Thi NBD, Lin C-Y, Kumar G (2016) Waste-to-wealth for valorization of food waste to hydrogen and methane towards creating a sustainable ideal source of bioenergy. J Clean Prod 22:29–41

96. Li Y, Khanal SK (2016) Bioenergy: principles and applications. Wiley

97. Bridgeman J (2012) Computational fluid dynamics modelling of sewage sludge mixing in an anaerobic digester. Adv Eng Softw 44:54–62

98. Conklin AS, Chapman T, Zahller JD, Stensel HD, Ferguson JF (2008) Monitoring the role of aceticlasts in anaerobic digestion: activity and capacity. Water Res 42:4895–4904

99. Halalsheh M, Kassab G, Yazajeen H, Qumsieh S, Field J (2011) Effect of increasing the surface area of primary sludge on anaerobic digestion at low temperature. Bioresour Technol 102:748–752

100. Kaparaju P, Buendia I, Ellegaard L, Angelidakia I (2008) Effects of mixing on methane production during thermophilic anaerobic digestion of manure: lab-scale and pilot-scale studies. Bioresour Technol 99:4919–4928

101. Suwannoppadol S, Ho G, Cord-Ruwisch R (2011) Rapid start-up of thermophilic anaerobic digestion with the turf fraction of MSW as inoculum. Bioresour Technol 102:7762–7767

102. Chandra R, Takeuchi H, Hasegawa T (2012) Methane production from lignocellulosic agricultural crop wastes: a review in context to second generation of biofuel production. Renew Sustain Energy Rev 16:1462–1476

103. Kobayashi T, Xu K-Q, Li Y-Y, Inamori Y (2012) Effect of sludge recirculation on characteristics of hydrogen production in a two-stage hydrogen–methane fermentation process treating food wastes. Int J Hydrog Energy 37:5602–5611

104. Zuo Z, Wu S, Zhang W, Dong R (2014) Performance of two-stage vegetable waste anaerobic digestion depending on varying recirculation rates. Bioresour Technol 162:266–272

105. Thamsiriroj T, Murphy JD (2011) Modelling mono-digestion of grass silage in a 2-stage CSTR anaerobic digester using ADM1. Bioresour Technol 102:948–959

106. Boe K, Angelidaki I (2009) Serial CSTR digester configuration for improving biogas production from manure. Water Res 43:166–172

107. Walker M, Banks CJ, Heaven S (2009) Two-stage anaerobic digestion of biodegradable municipal solid waste using a rotating drum mesh filter bioreactor and anaerobic filter. Bioresour Technol 100:4121–4126

108. Mshandete A, Murto M, Kivaisi AK, Rubindamayugi MST, Mattiasson B (2004) Influence of recirculation flow rate on the performance of anaerobic packed-bed bioreactors treating potato-waste leachate. Environ Technol 25:929–936

109. Xie S, Hai FI, Zhan X, Guo W, Ngo HH, Price WE, Nghiem LD (2016) Anaerobic co-digestion: a critical review of mathematical modelling for performance optimization. Bioresour Technol 222:498–512

110. Yu L, Zhao Q, Ma J, Frear C, Chen S (2012) Experimental and modeling study of a two-stage pilot scale high solid anaerobic digester system. Bioresour Technol 124:8–17

111. Padmasiri SI, Zhang J, Fitch M, Norddahl B, Morgenroth E, Raskin L (2007) Methanogenic population dynamics and performance of an anaerobic membrane bioreactor (AnMBR) treating swine manure under high shear conditions. Water Res 41:134–144

112. Palatsi J, Viñas M, Guivernau M, Fernandez B, Flotats X (2011) Anaerobic digestion of slaughterhouse waste: main process limitations and microbial community interactions. Bioresour Technol 102:2219–2227

113. Fuess LT, Kiyuna LSM, Júnior ADNF, Persinoti GF, Squina FM, Garcia ML, Zaiat M (2017) Thermophilic two-phase anaerobic digestion using an innovative fixed-bed reactor for enhanced organic matter removal and bioenergy recovery from sugarcane vinasse. Appl Energy 189:480–491

114. Zhang D, Zhu W, Tang C, Suo Y, Gao L, Yuan X, Wang X, Cui Z (2012) Bioreactor performance and methanogenic population dynamics in a low-temperature (5–18 °C) anaerobic fixed-bed reactor. Bioresour Technol 04:136–143

115. Ráduly B, Gyenge L, Szilveszter S, Kedves A, Crognale S (2016) Treatment of corn ethanol distillery wastewater using two-stage anaerobic digestion. Water Sci Technol 74:431–437

116. Jang HM, Ha JH, Park JM, Kim M-S, Sommer SG (2015) Comprehensive microbial analysis of combined mesophilic anaerobic–thermophilic aerobic process treating high-strength food wastewater. Water Res 73:291–303

117. Li W-W, Yu H-Q (2011) Physicochemical characteristics of anaerobic H2-producing granular sludge. Bioresour Technol 102:8653–8660

118. Li R, Chen S, Li X (2010) Biogas production from anaerobic co-digestion of food waste with dairy manure in a two-phase digestion system. Appl Biochem Biotechnol 160:643–654

119. Wu S, Dang Y, Qiu B, Liu Z, Sun D (2015) Effective treatment of fermentation wastewater containing high concentration of sulfate by two-stage expanded granular sludge bed reactors. Int Biodeterior Biodegrad 104:15–20

120. Zahedi S, Solera R, Micolucci F, Cavinato C, Bolzonella D (2016) Changes in microbial community during hydrogen and methane production in two-stage thermophilic anaerobic co-digestion process from biowaste. Waste Manag 49:40–46

121. Wang X, Zhao Y (2009) A bench scale study of fermentative hydrogen and methane production from food waste in integrated two-stage process. Int J Hydrog Energy 34:245–254

122. Campo G, Cerutti A, Zanetti M, Scibilia G, Lorenzi E, Ruffino B (2018) Enhancement of waste activated sludge (WAS) anaerobic digestion by means of pre-and intermediate treatments. Technical and economic analysis at a full-scale WWTP. J Environ Manag 216:372–382

123. Bouallagui H, Lahdheb H, Ben RE, Rachdi B, Hamdi M (2009) Improvement of fruit and vegetable waste anaerobic digestion performance and stability with co-substrates addition. J Environ Manag 90:1844–1849

124. Cavinato C, Bolzonella D, Pavan P, Fatone F, Cecchi F (2013) Mesophilic and thermophilic anaerobic co-digestion of waste activated sludge and source sorted biowaste in pilot-and full-scale reactors. Renew Energy 55:260–265

125. Lin J, Zuo J, Gan L, Li P, Liu F, Wang K, Chen L, Gan H (2011) Effects of mixture ratio on anaerobic co-digestion with fruit and vegetable waste and food waste of China. J Environ Sci (China) 23:1403–1408

126. Lee D-Y, Ebie Y, Xu K-Q, Li Y-Y, Inamori Y (2010) Continuous H2 and CH4 production from high-solid food waste in the two-stage thermophilic fermentation process with the recirculation of digester sludge. Bioresour Technol 101:S42–S47

127. Koutrouli EC, Kalfas H, Gavala HN, Skiadas IV, Stamatelatou K, Lyberatos G (2009) Hydrogen and methane production through two-stage mesophilic anaerobic digestion of olive pulp. Bioresour Technol 100:3718–3723

128. Hidalgo D, Martín-Marroquín JM (2015) Biochemical methane potential of livestock and agri-food waste streams in the Castilla y León Region (Spain). Food Res Int 73:226–233

129. Nasr N, Elbeshbishy E, Hafez H, Nakhla G, El Naggar MH (2012) Comparative assessment of single-stage and two-stage anaerobic digestion for the treatment of thin stillage. Bioresour Technol 111:122–126

130. Yu L, Ma J, Chen S (2011) Numerical simulation of mechanical mixing in high solid anaerobic digester. Bioresour Technol 102:1012–1018

131. Saeman JF (1945) Kinetics of wood saccharification—hydrolysis of cellulose and decomposition of sugars in dilute acid at high temperature. Ind Eng Chem 37:43–52

132. Liu X, Li R, Ji M, Han L (2013) Hydrogen and methane production by co-digestion of waste activated sludge and food waste in the two-stage fermentation process: substrate conversion and energy yield. Bioresour Technol 146:317–323

133. Luo G, Xie L, Zhou Q, Angelidaki I (2011) Enhancement of bioenergy production from organic wastes by two-stage anaerobic hydrogen and methane production process. Bioresour Technol 102:8700–8706

134. Borowski S (2015) Temperature-phased anaerobic digestion of the hydromechanically separated organic fraction of municipal solid waste with sewage sludge. Int Biodeterior Biodegrad 105:106–113

135. Akgul D, Cella MA, Eskicioglu C (2017) Influences of low-energy input microwave and ultrasonic pretreatments on single-stage and temperature-phased anaerobic digestion (TPAD) of municipal wastewater sludge. Energy 123:271–282

136. Astals S, Musenze RS, Bai X, Tannock S, Tait S, Pratt S, Jensen PD (2015) Anaerobic co-digestion of pig manure and algae: impact of intracellular algal products recovery on co-digestion performance. Bioresour Technol 181:97–104

Chapter 10
Biofuel Production

Industrial production of biofuels

Although AD is a sophisticated and widely applied technique, it faces many technical, economic, and social challenges. A few of these are discussed below.

10.1 Comparative Analysis of Single- and Two-Stage Processes

Prevailing digesters employed for the AD encompasses chiefly a 1S-AD or 2S-AD bioreactors. 1S-AD in which hydrolytic, acetogenesis, and methanogenesis reactions takes place concurrently emerges as a prominent configuration of digester [1]. In a study [2] it is found that the production of biogas in 1S-wet AD with substrate loading from 3.7 to 12.9 g-VS/L·d, and actualized greater biomethane production and solids removal on increment in the substrate load to 9.2 g-VS/L·d. However, the operation of these reactors at an elevated substrate loading and for highly organic-rich biowaste was considered as difficult since these wastes endures hasty acidogenesis causing hindrances of methanogen activity [3]. The maximal substrate load suggested for 1S-AD of organic waste was limited to 3.6 kg VS/m^3 d [4]. 2S-AD possess the benefit of buffering substrate load rate in stage 1, letting a fixed substrate loading in stage 2 methane producing bioreactor [5]. Many researches culminated about the stability of 2S-AD being proficient when compared to 1S-AD. 1S and 2S-AD of FVW and food waste mixture was correlated for system stability. 2S-AD offered a stability condition during greater substrate loading and finer buffering effectiveness when compared to 1S-AD, thus affording greater biodegradation and profits. Greater substrate loading indicates greater working potential for similar volume of reactor that can fetch exceptional advantages economically. It is suggested that [6] a 2S-codigestion is opt-in comparison to 1S-AD for treating lipid rich wastes. A study [7] the two and single stage systems, on the basis of bioenergy recovery and concluded

© Springer Nature Singapore Pte Ltd. 2022
K. Sudalyandi and R. Jeyakumar, *Biofuel Production Using Anaerobic Digestion*,
Green Energy and Technology, https://doi.org/10.1007/978-981-19-3743-9_10

that 18.5% increment in overall energy production through 2S-AD. The average gain in specific methane yield (SMY) in two-stage digestion was 21–23% when compared 1S AD [416]. This is consistent with other studies [8, 9] explaining improved methane production using biowastes as substrates in comparison to 1S-AD. Comparing with 1S and 2S-AD (both in thermophilic and mesophilic condition), TPAD, was found to be advantageous in magnifying the reduction of solids and pathogens, heightening the biogas yield, improving the performance of bioreactor and also shortening of HRT. In the meantime, the reduction of pathogens in TPAD surpassed single-stage mesophilic anaerobic digestion [10, 11] 0.1S-AD and TPAD treating lipid rich organic waste with and without recirculation were performed and their stability were assessed [12]. Comparable solids and COD were biodegraded in 1S-AD and 2S-AD without recirculation. Recycling was tested to be potent in the two-stage system and it converts the unsaturated long-chain fatty acids into saturated ones Fig. 10.1 shows a schematic representation of comparison between single and two-stage AD.

10.2   Performance Enhancement

The accessibility and easily biodegradable potential of the various feedstocks are essential to enhance the methane yield through codigesting process [13]. The anaerobic co-digestion (AcoD) technology help in improved nutrients equilibrium in the bioreactors. Incessant investigations are reporting about the codigesting potential of pig waste and microalgae, milk industry effluent and cattle waste [14], sludge biosolids and glycerol [15] at disparate operational environment that substantiates enrichment in the methane yield as to the single substrate feed. Present research on co-digestion focuses on recognition of novel probable feedstocks, the appropriate codigesting substrate proportion and operational environments such as inoculating proportion, substrate loading, and temperature [16]. Lately, a research [17] revealed the disintegration of dual substrates (wheat straw and cattle waste) during codigestion by applying H_2O_2 (moistured form) bolstered up the biodegradation and biogas production. In codigestion, the methanogens shuffled from acetoclastic to hydrogenotrophic methanogens predominantly available in agricultural wastes [18]. Thus, this method may likely downturn VFA concentration and create a favorable condition thereby escalating organics degradation to enhance the methane yield. Effluent recirculation from digestion reactor was adapted for neutralization of treated acidic residue to augment biogas yield [19]. Thus, to overwhelm the risks and difficulties of codigestion, designing of suitable digesters, emerging process description, classifying substrates on the basis of biodegrading potential, and availability are essential.

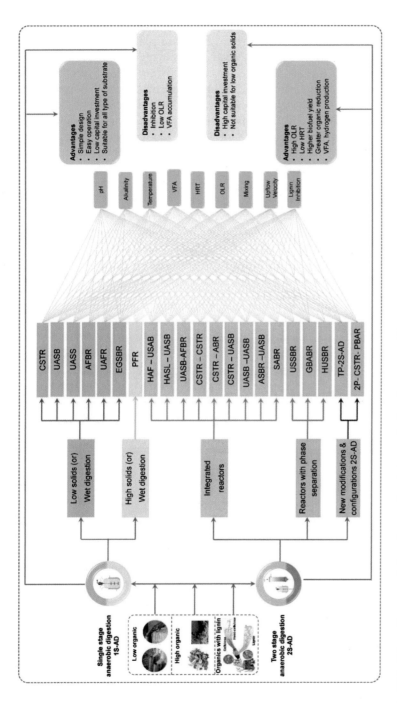

Fig. 10.1 Schematic representation of comparison between single and two-stage AD

10.2.1 Inhibition Mitigation

The biodegradation of substrates may be reduced or suppressed owing to the existence of inhibiting materials [20]. Inadequacy of essential micro nutrients as stated in research findings perturbs the functions and activities of the hydrolytic bacteria and key enzymes, and transforms the ecological circumstances for development of microbes [21], coaxing system letdown. Adding the trace of elements demarcated suppression of organic acids build up, and stabilize. In many findings, whereas addition of one type of trace element alone, e.g. Cobalt, Molybdenum, or Nickel, bettered the stability of reactor only for little time. The Selenium and Cobalt were highly essential in which 0.16 and 0.22 mg/kg of Se and Co concentration were found to be optimal as suggested by a study [22] And the study found that authors have added that these elements can proficiently ebb organic acids. Zeolite, and an eco-friendly element is employed to amend the stability anaerobic codigestion [23]. Zeolite was extensively employed as ion exchanging materials for removing ammonium and absorbing heavy metals due to its conducting characteristics and as a result it can be employed as glues that are toxic for microbes in anaerobic codigestion [24]. The impact of zeolite on the overall co digestion of fecal matter and food wastes was explored. It is found that the methane generation increased from 44.10 to 65.3%. Inhibition of propionic acid may happen in certain bioreactors while acetogenic microbes are comparatively feeble in development and sulphate can be added to alleviate inhibition of propionic acid.

10.3 Foaming

Foaming issues transpires in wet digestion due to the accumulation of surfactants that may originate from substrates [25]. The foam formation is elicited via abrupt discharge of gases. Owing to the vast variations in methane and carbondioxide liquefaction. i.e. eventual CH_4 exist in gas phase of digested residue in the form of bubbles, whereas a considerable concentration of carbon dioxide is liquefied in aqueous phase as carbonates and carbonic acid based on operational conditions [26]. Several approaches were established and employed for controlling foam and employed recently. Applying antifoam agents has been developed as an extensive approach for controlling foam. Foaming problems evoked through abrupt and repeated passage of gas discharge and could be simply controlled through application of antifoaming additives [27]. Multifarious compounds which includes vegetable oils, salts, and alcoholic byproducts, could be employed as antifoaming additives [25]. Antifoaming additives such as rapeseed oil (0.05%) and octanoic acid (0.5%) has been employed in a study investigating batch and continuous digesters. These additives have been observed as more effective when compared to other antifoaming additive tributyl phosphate in controlling foam [25]. Natural oils and esters of fatty acids and other marketable antifoaming additives could be employed for reactors that have foaming issues for a longer period of time [27]. The foaming problems which are incited through excess substrate loading, then the loading rate of non-biodegradable feedstocks must be strictly regulated, particularly while digesting non- biodegradable substrates that are rich in carbohydrates and starch.

10.4 The Latest Anaerobic Digestion Systems and Technologies

Improvement in reactor configurations and operational approaches is a significant feature to boost the substrate loading, biogas production, and stable performance of reactors. A compacted 3 phase reactor was used for treating food wastes[28]. Where in 3 separate compartments for hydrolytic, acidogenic, and methanogenesis phase integrated into a self-regulating compartment, and attained greater biogas percentage of 24–54% [29]. A study [30], suggested using of acidogenic phase for reduction of sulphate to hydrogen sulphide as sulphate reducers, possess surpassing development when compared to methanogenic microbes and can endure acidification environment [31]. A TPAD (phase 1 working on thermophilic and phase 2 working on mesophilic) was used for treating oil rich kitchen waste. Since thermophilic condition affirms to improve hydrolytic phase and let elevated feeding of lipids [12], whereas mesophilic condition favors methanogenic phase. Installing equipment to recirculate digested effluent clout the microbes and essential sources in digested residue and deflates residue discharge. Besides, effluent recirculation take place a vital part in stream mixing, substrate dilution, and it is considered as the more feasible and efficient bioenergy producing process. As well, biogas upgrading could metamorphose CH_4 into a blend of syngas, hydrogen and carbonmonoxide via various reforming processes [32]. Furthermore, H_2 production intensifies the energetic rate of biogas and the consumption ratio in electrical energy producing fuel cells.

10.5 Economics and Scale-Up of 1-SD

An energy balance and economic analysis is essential for IS-AD to be implemented at large scale. It helps stakeholders and policy makers to choose the appropriate and efficient technology. The economic and scaling up of a process can be done by taking into consideration of following parameters such as input energy, heat, power production and financial exploration.

For energy analysis in 1S-AD, it is have considered that heat and electricity requirements as input energy and methane yield as output energy [33]. The input electricity include the electricity utilized for pumping (substrate and digested liquor) and for mixing. The input heat comprises the heat energy needed to increase the

temperature of substrate to the temperature of digester (in case of thermophilic digestion) and to compensate the loss of heat through walls, floor and digester cover. In
another study, further additional parameters such as the energy spent for dewatering,
conveyance and the land fill after dewatering included in later studies [34].

The following equation helps in arriving values for total input energy.

The energy spent towards electricity and heat can be calculated as follows:

$$E_{elect} = Q_1\theta_1 + \text{Vol} * \omega_1/Q_1 \cdot VS \tag{10.1}$$

where E_{elec} is the spent electricity power (kJ/gVS); Q_1 is the flow rate of influent;
VS is the volatile solids of feedstock fed into the digester; Vol is the reactor volume
(m^3); θ_1 is the electrical power utilized for pumping (1800 kJ/m^3); ω_1 is the electrical
power utilized for stirring ($300 \text{ kJ/m}^3_{digester}$ d);

$$E_{heat} = E_{heat, rt} + E_{heat, cl} \tag{10.2}$$

where E_{heat} is the spent heat energy (kJ/g VS); E_{heat},rt is the spent heat energy towards
influent temperature raise to temperature of digester; E_{heat},cl is the heat energy spent
towards compensation of heat loss of reactor walls.

$$E_{heat, rt} = \rho_1 * Q_1 * \gamma_1 * (T_{diges} - T_{inf})/Q_1 \cdot VS \tag{10.3}$$

where ρ_1 is the influent density (1000 kg/m^3).
$\gamma 1$ is the influent specific heat (4.18 kJ/kg °C).
T_{diges} is the temperature of anaerobic digester.
T_{inf} is the influent temperature (°C)

$E_{heat, cl}$
$$= (SA_w * (T_{diges}T_{ai}) * k_{Hw} + SA_{fl} * (T_{diges} - T_{ea}) * k_{fl} + SA_{dc} * (T_{diges}T_{ai}) * k_{dc})$$
$$* 3600 * 24 \tag{10.4}$$

where SA_w is the digester wall surface area (m^2); T_{ai} is the air temperature (°C);
k_{Hw} is the heat transfer coefficient of walls and lining of reactor ($0.8 \text{ W/m}^2 \text{ °C}$); SA_{fl}
reactor floor surface area (m^2); T_{ea} is the earth temperature k_{fl} is the heat transfer
coefficient of digester floor; SA_{dc} is the digester cover surface area; k_{dc} is the heat
transfer coefficient of reactor cover.

The recovery of heat energy can be calculated as follows:

$$E_{heat.rec} = \rho_1 * Q_1 * \gamma_1 * (T_{diges} - T_{inf}) * \phi \tag{10.5}$$

where

$E_{\text{heat.rec}}$ is the recovered heat energy (kJ/g VS).
ϕ: recovered heat percentage (80%).
The output energy in single stage digestion can be calculated as follows:

$$E_{ou} = M_y * Vol * \xi_{met} * \eta_{met} / Q_1 * VS \tag{10.6}$$

where E_{ou} is the energy output (kJ/g VS).
M_y is the yield of CH_4 (m^3/m^3/reactor.d).
ξ_{met} is the lower heating value of CH_4 (35,800 kJ/m^3).
η_{me} is the factor energy conversion efficiency of CH_4 (0.9).
The energy balance (the energy spent must be balanced by the energy output) can be calculated as follows:

$$\Delta E = E_{ou} - (E_{elect} + E_{heat} - E_{heat.rec}) \tag{10.7}$$

The energy spent towards dewatering using filter press can be calculated as follows:

$$E_{dew} = QN_{TDS} * SE_{TDS} * 3600 \tag{10.8}$$

where E_{dew} is the energy spent towards dewatering of digested residue (kJ/d).
QN_{TDS} is the amount of dry solid of digested residue (kg).
SE_{TDS} is the energy consumed for dry solid of digestate (35×10^{-3} kWh/kg) [558].
The energy spent towards transport of digestate through semi-trailer vehicles and land application of digestate through spreader and both can be calculated as follows:

$$E_{ddt} = QN_{dd} * \text{dist} * 2 * FC_v * \xi_F / t_{cap} \tag{10.9}$$

where E_{ddt} is the energy spent towards transportation of dewatered digested residue (kJ/d).
QN_{dd} is the amount of dewatered digested residue (kg/d).
dist is the distance between the site and the place of land filling.
fv: the fuel consumed by automobile.
ξ_F: the fuel (diesel) heating value (39×10^3 kJ/L diesel).
t_{cap} is the vehicles conveyance capability (14.1×10^3 kg) [559].

$$E_{LA} = QN_{TDS} * SE_{TDS} * 3600 \tag{10.10}$$

where E_{LA} is the energy spent towards land application of digestate (kJ/d).
The energy ratio of single stage digestion is described as the ratio of output energy to input energy and it was calculated as:

$$ER = Eo/Ei \qquad\qquad (10.11)$$

where ER is the energy ratio,
   Eo is the output energy (kJ/d).
   Ei is the input energy (kJ/d).

10.6   Economics and Scale-Up 2-AD

Solids reduction price and profits achieving via electric power selling are the two aspects which have mammoth influence on ultimate profit, hence, should be appraised and quantified while gauging the appositeness of these solutions in real situations. A comparative assessment of techno, economic and environmental aspects was done for 2S-AD and 1S-AD treating primary and secondary sludge [35]. Greater benefits obtained economically is probably attained by progressing the effluent treatment plant with post-aerobic digestion phase. Here, in this study, the cost analysis was not transacted on actual, effluent treatment plant in specific, the capital cost for progressing the prevailing treatment plant with the newer instruments. A study [36] that have performed a technical and economic analysis of prior and transitional lytic disintegration in a pilot scale treatment plant (2,000,000 p.e.), revealed that influx of heat or combine pretreatments can elevate incomes from selling of electrical power of 13–25% as to no lysis treatments. The capital cost investment for hydrolysis digesters were estimated by Leite et al. [38] to check the economic feasibility of their large scale investigation conducted on 1S and 2S- AD for treating waste activated sludge. The investments achieved from the 2S-AD with regard to 1S-AS were 95,451£/yr. Around 35 percent of the incomes was associated with increase of electricity selling (33,458 £/yr) whereas the 65% of profits were achieved via sludge disposal (around 620 tons/yr). The payback period was found to be 3 years. In view of the large scale outcome on the stability of 1S-AD and 2S-AD hinged on mass and energy equilibrium, and costs [38]. The 2S-AD obtained a net energy generation of 55 percent greater than 1S-AD. A simple economic assessment was conducted in an investigation [39], contrasting the 2S-AD process to alcoholic fermentation process to test the potentiality of molasses. It was estimated that the cost benefit of the biogas generated (0.206 $/L-molasses/d) and this is slightly greater when compared to ethanol produced (0.196 $/L-molasses/d) in addition to slight increase in energy recovery.

10.7   Comparison of Energy and Economic Analysis of Single and Two-Stage AD

Energy ratio in excess of 1 infers net energy generation. Table 10.1 shows the energy and economic analysis of single and two-stage digester in reported in various literature. In some studies, as indicated in Table 10.1 the researchers have contradictorily reported too high energy ratio. For example, in a study energy ratio of 53.2 was achieved and that was found to be too high value for energy ratio. Mesophilic single stage digesters are more commonly employed than thermophilic single stage digester due to less energy requirement and economic viability. The obtainability of residual heat (the heat energy that remains even after the heat demand of digester is subtracted from total heat generation from CHP unit) and the electrical energy were frequently greater in mesophilic single stage digestion than thermophilic digestion [37]. For example, 20 kcal/d * 10^5 of residual heat energy and 40 kcal/d * 10^5 of electrical energy have been obtained in single stage thermophilic digestion [37]. At the same time a higher residual energy and electrical energy of 28 kcal/d * 10^5 and 55 kcal/d * 10^5 through mesophilic single stage digestion. Also it is reported that a higher net energy yield (17.40 kJ/g VS) in mesophilic single stage digestion when compared to thermophilic single stage digestion (17.29 kJ/g VS) [414]. The efficiency of mesophilic single stage digestion can be increased by thickening the feed material by dewatering and improving the biodegradability of the feed materials by pretreating [40–46].

The investment cost of 1S- AD depends upon the digester type (i.e. specific requirement of installation). The entire cost of investment of a digester could differ from 2500 to 7500 euro [62]. The components of a digester could get malfunctioned during operation. In such cases, they must be repaired. Accordingly, periodical maintenance of digester is needed. The investment cost for maintenance in single stage farm digester was 2500 euro per 2000 operation hours [62]. Concerning the substrate or feedstock cost, cost of transportation should be considered.

The economic feasibility study of thermophilic 1S-AD treating food waste at pilot scale level, reveals that in waste to energy process, collection of gate fee is mandatory [38]. Gate fee covers cost associated with operation, labor and disposal process. The gate fee is the cost charged for waste (substrate) used for digestion. Recently the gate fee is charged to be 85 euro/T waste [37]. If the digestate sent for composting it has a cost of 60 euro/T waste, and a net profit of 255,877 euro/year could be obtained through 1S-AD [38].

Working on thermophilic 1S-AD treating waste activated sludge (WAS), a net profit of 147,375 euro/year [37], a net profit (electricity) of 178, 265 euro/year for single stage mesophilic and 157,043 euro/year for single stage thermophilic digestion. For a mesophilic single stage digester treating sludge, a net profit of 32,717 euro/year obtained [63].

Electrical power for running 1S-AD can be obtained from combined heat and power (CHP) unit of biogas plant. No extra cost will be spent if the required electric power was fulfilled through the CHP unit. Upon analyzing energy generation in field

Table 10.1 Comparative energy analysis of various 1S-AD and 2S-AD

S. No	Type of digester	Mode of temperature	Substrate	Input energy (kWh)	Output energy (kWh)	Net energy production	Energy ratio	Net profit (USD)	References
Single stage anaerobic digester (1S-AD)									
1	CSTR	Thermophilic	Food waste	0.0006888889[b]	0.0046861111[b]	0.003997[b]	6.80	0.00092*	[47]
2	CSTR	Thermophilic	Food waste	390.55	4016.3	3758.48	15.58	864.45*	[48]
3	CSTR	Mesophilic	Food waste	234.33	3877.9	3712.01	23.37	853.7623*	[48]
4	CSTR	Thermophilic	Waste activated sludge	1114.57[c]	2560[d]	1445.43	2.29	332.4489*	[49]
5	CSTR	Thermophilic	Organic fraction of municipal solid waste	2.09[b]	5.207[b]	3.117[b]	2.49	0.716*	[50]
6	Solid state anaerobic digester	Mesophilic	Dewatered sludge	673.77	570	−103.77	−0.85	−23.87*	[51]
7	Single stage-solid state anaerobic digester	Mesophilic	Dairy manure, corn stover and tomato residues	1183.3[a]	62,988.9[a]	61,805.6[a]	53.2	14,215.3*	[52]
8	CSTR	Thermophilic	Food waste	40.19	43	2.81	1.07	0.646	[53]
9	CSTR	Mesophilic	Municipal solid waste	33.45	35.40	1.95	0.42	0.45	[54]

Table 10.1 (continued)

S. No	Type of digester	Mode of temperature	Substrate	Input energy (kWh)	Output energy (kWh)	Net energy production	Energy ratio	Net profit (USD)	References
10	Single stage digester	Mesophilic	Waste activated sludge	696.45	770	73.55	1.10	16.917	[55]
11	CSTR	Thermophilic	Cassava stillage	58.06	78.72	20.66	1.36	4.75	[56]
Two-stage anaerobic digester (2S-AD)									
12	Two-stage CSTR	Thermophilic	Food waste	0.0001527778[b]	0.0047055556[b]	0.004553[b]	30.8	0.001047*	[47]
13	Two-stage CSTR	Thermophilic/Mesophilic	Food waste	390.55	3819.7	3560.09	14.72	818.8207*	[48]
14	Two-stage CSTR	Thermophilic	Waste activated sludge	1299.087[c]	3000[d]	1700.913	2.30	391.21*	[49]
15	Two-stage CSTR	Thermophilic	Biowaste	4.1467[b]	16.57[b]	12.4233[b]	4	2.857*	[57]
16	Two-stage batch reactor	Thermophilic	Cassava stillage	58.06	9.269	−48.791	0.16	−11.22	[56]
17	Two-stage CSTR	Thermophilic	Food waste	4.147[b]	21.89[b]	17.743[b]	5.28	4.08*	[58]

(continued)

Table 10.1 (continued)

S. No	Type of digester	Mode of temperature	Substrate	Input energy (kWh)	Output energy (kWh)	Net energy production	Energy ratio	Net profit (USD)	References
18	Semi continuous two-stage anaerobic digester	Thermophilic	Waste activated sludge	1183	6365	5182	5.38	1191.86	[59]
19	Two-stage batch reactor	Mesophilic	Organic market waste	29.028	139.77	110.74	4.82	25.47	[60]
20	Two-stage batch reactor	Thermophilic	House solid waste	69.67	12.14	−57.53	0.17	−13.23	[61]

[a] 1 GJ is equal to 277.778 kWh, [b] 1 kJ is equal to 0.000277778 kWh, [c] 1 kcal is equal to 0.00116222 kWh, [d] 1 MWh is equal to 1000 kWh, *calculated

scale 1S-AD. The generated gas could be utilized to produce electrical power and heat via CHP generation unit [62]. The produced electricity and heat can be used by nearby communities. Some portion of produced heat can be utilized in the process regulation and in some cases, for substrate sterilization or pretreatment if required.

What Next?

The practicability and viability of anaerobic codigestion for methane generation was flourishing in recent years. This pinpoints the efficiency, in addition to upsurge the methane yield through codigesting various substrates. But, some of the offbeat tasks that need to be accosted for upgrading the codigestion to large scale and industrial extent. Numerous processes should be upheld which includes buffering effect, the generation extent of methane, nutrients equilibrium, and stable microbial growth. Application of dual substrates in pretreatment and hydrolytic phases is yet to be probed. Application of nanoparticles can be the best option to screen and regulate the system. Attempts are made to recede the inhibitory effects of NH_4, organic acids, and sulphides [64, 65]. Amalgamation of different waste materials [66] and their conjugate influence on co-digestion can be substantiated through continuous reactor [67].

Contemporary and universal pretreatment approaches can be applied in future. Assessment of cost and environment aspects must be performed eventually for choosing favorable disintegration. Efficient techniques to eradicate foam forming agents and to remove the foams using enzymes or live cultures needs to be investigated. An in-depth investigation of biological steps of digestion system and developments in metabolic engineering are essential to enhance the stable performance of reactors.

References

1. Dalkılıc K, Ugurlu A (2015) Biogas production from chicken manure at different organic loading rates in a mesophilic-thermopilic two stage anaerobic system. J Biosci Bioeng 120:315–322
2. Fu X, Hu Y (2016) Comparison of reactor configurations for biogas production from rapeseed straw. BioResources 11:9970–9985
3. Jensen PD, Astals S, Lu Y, Devadas M, Batstone DJ (2014) Anaerobic codigestion of sewage sludge and glycerol, focusing on process kinetics, microbial dynamics and sludge dewaterability. Water Res 67:355–366
4. Song Z, Zhang C (2015) Anaerobic codigestion of pretreated wheat straw with cattle manure and analysis of the microbial community. Bioresour Technol 86:128–135
5. Kallistova AY, Goel G, Nozhevnikova AN (2014) Microbial diversity of methanogenic communities in the systems for anaerobic treatment of organic waste. Microbiology 83:462–483
6. Li Q, Li Y-Y, Qiao W, Wang X, Takayanagi K (2015) Sulfate addition as an effective method to improve methane fermentation performance and propionate degradation in thermophilic anaerobic co-digestion of coffee grounds, milk and waste activated sludge with AnMBR. Bioresour Technol 185:308–315
7. Abbasi T, Ramasamy EV, Khan FI, Abbasi SA (2012) Regional EIA and risk assessment in a fast developing country

8. Banks CJ, Zhang Y, Jiang Y, Heaven S (2012) Trace element requirements for stable food waste digestion at elevated ammonia concentrations. Bioresour Technol 104:127–135

9. Montalvo S, Guerrero L, Borja R, Sánchez E, Milán Z, Cortés I, Angeles de la la Rubia M (2012) Application of natural zeolites in anaerobic digestion processes: a review. Appl Clay Sci 58:125–133

10. Wang X, Zhang L, Xi B, Sun W, Xia X, Zhu C, He X, Li M, Yang T, Wang P, Zhang Z (2015) Biogas production improvement and C/N control by natural clinoptilolite addition into anaerobic co-digestion of Phragmites australis, feces and kitchen waste. Bioresour Technol 180:192–199

11. Kougias PG, Boe K, Tsapekos P, Angelidaki I (2014) Foam suppression in overloaded manure-based biogas reactors using antifoaming agents. Bioresour Technol 153:198–205

12. Fontana A, Campanaro S, Treu L, Kougias PG, Cappa F, Morelli L, Angelidaki I (2018) Performance and genome-centric metagenomics of thermophilic single and two-stage anaerobic digesters treating cheese wastes. Water Res 134:181–191

13. Hagos K, Zong J, Li D, Liu C, Lu X (2017) Anaerobic co-digestion process for biogas production: progress, challenges and perspectives. Renew Sustain Energy Rev 76:1485–1496

14. Lindorfer H, Demmig C (2016) Foam formation in biogas plants–a survey on causes and control strategies. Chem Eng Technol 39:620–626

15. Yan BH, Selvam A, Wong JWC (2016) Innovative method for increased methane recovery from two-phase anaerobic digestion of food waste through reutilization of acidogenic off-gas in methanogenic reactor. Bioresour Technol 217:3–9

16. Lau CS, Tsolakis A, Wyszynski ML (2011) Biogas upgrade to syn-gas (H2–CO) via dry and oxidative reforming. Int J Hydrogen Energy 36:397–404

17. Zhang Q, Hu J, Lee D-J (2016) Biogas from anaerobic digestion processes: research updates. Renew Energy 98:108–119

18. Tasnim F, Iqbal SA, Chowdhury AR (2017) Biogas production from anaerobic co-digestion of cow manure with kitchen waste and water hyacinth. Renew Energy 109:434–439

19. Xie S, Wickham R, Nghiem LD (2017) Synergistic effect from anaerobic co-digestion of sewage sludge and organic wastes. Int Biodeterior Biodegradation 116:191–197

20. Wan S, Sun L, Sun J, Luo W (2013) Biogas production and microbial community change during the Co-digestion of food waste with chinese silver grass in a single-stage anaerobic reactor. Biotechnol Bioprocess Eng 18:1022–1030

21. Romero-Güiza MS, Vila J, Mata-Alvarez J, Chimenos JM, Astals S (2016) The role of additives on anaerobic digestion: a review. Renew Sustain Energy Rev 58:1486–1499

22. Song Y-C, Kwon S-J, Woo J-H (2004) Mesophilic and thermophilic temperature co-phase anaerobic digestion compared with single-stage mesophilic-and thermophilic digestion of sewage sludge. Water Res 38:1653–1662

23. Chamy R, Vivanco E, Ramos C (2011) Anaerobic mono-digestion of Turkey manure: efficient revaluation to obtain methane and soil conditioner. J Water Resour Prot 3:584

24. Liu X, Wang W, Shi Y, Zheng L, Gao X, Qiao W, Zhou Y (2012) Pilot-scale anaerobic co-digestion of municipal biomass waste and waste activated sludge in China: effect of organic loading rate. Waste Manag 32:2056–2060

25. Moon HC, Song IS (2011) Enzymatic hydrolysis of foodwaste and methane production using UASB bioreactor. Int J Green Energy 8:361–371

26. Agyeman FO, Tao W (2014) Anaerobic co-digestion of food waste and dairy manure: effects of food waste particle size and organic loading rate. J Environ Manage 133:268–274

27. Zhang C, Xiao G, Peng L, Su H, Tan T (2013) The anaerobic co-digestion of food waste and cattle manure. Bioresour Technol 129:170–176

28. Hafez H, Nakhla G, El Naggar H (2010) An integrated system for hydrogen and methane production during landfill leachate treatment. Int J Hydrogen Energy 35:5010–5014

29. Aslam M, McCarty PL, Shin C, Bae J, Kim J (2017) Low energy single-staged anaerobic fluidized bed ceramic membrane bioreactor (AFCMBR) for wastewater treatment. Bioresour Technol 240:33–41

30. Dareioti MA, Kornaros M (2014) Effect of hydraulic retention time (HRT) on the anaerobic co-digestion of agro-industrial wastes in a two-stage CSTR system. Bioresour Technol 167:407–415

31. Conklin AS, Chapman T, Zahller JD, Stensel HD, Ferguson JF (2008) Monitoring the role of aceticlasts in anaerobic digestion: activity and capacity. Water Res 42:4895–4904

32. Pan X, Xie D, Yu RW, Lam D, Saddler JN (2007) Pretreatment of Lodgepole pine killed by Mountain Pine beetle using the ethanol organosolv process: fractionation and process optimization. Ind Eng Chem Res 46:2609–2617

33. Xiao B, Qin Y, Wu J, Chen H, Yu P, Liu J, You Li Y (2018) Comparison of single-stage and two-stage thermophilic anaerobic digestion of food waste: performance, energy balance and reaction process. Energy Convers Manag 156:215–223

34. Xiao B, Qin Y, Zhang W, Wu J, Qiang H, Liu J, You Li Y (2018) Temperature-phased anaerobic digestion of food waste: a comparison with single-stage digestions based on performance and energy balance. Bioresour Technol 249:826–834

35. Nualsri C, Kongjan P, Reungsang A (2016) Direct integration of CSTR-UASB reactors for two-stage hydrogen and methane production from sugarcane syrup. Int J Hydrogen Energy 41:17884–17895

36. Toumi J, Miladi B, Farhat A, Nouira S, Hamdi M, Gtari M, Bouallagui H (2015) Microbial ecology overview during anaerobic codigestion of dairy wastewater and cattle manure and use in agriculture of obtained bio-fertilisers. Bioresour Technol 198:141–149

37. Leite WRM, Gottardo M, Pavan P, Belli Filho P, Bolzonella D (2016) Performance and energy aspects of single and two phase thermophilic anaerobic digestion of waste activated sludge. Renew Energy 86:1324–1331

38. Ramakrishnan A, Surampalli RY (2012) Comparative performance of UASB and anaerobic hybrid reactors for the treatment of complex phenolic wastewater. Bioresour Technol 123:352–359

39. Jeihanipour A, Aslanzadeh S, Rajendran K, Balasubramanian G, Taherzadeh M (2013) High-rate biogas production from waste textiles using a two-stage process. Renew Energy 52:128–135

40. De Gioannis G, Diaz LF, Muntoni A, Pisanu A (2008) Two-phase anaerobic digestion within a solid waste/wastewater integrated management system. Waste Manag 28:1801–1808

41. Guwy AJ, Dinsdale RM, Kim JR, Massanet-Nicolau J, Premier G (2011) Fermentative biohydrogen production systems integration. Bioresour Technol 102:8534–8542

42. Zuo Z, Wu S, Zhang W, Dong R (2013) Effects of organic loading rate and effluent recirculation on the performance of two-stage anaerobic digestion of vegetable waste. Bioresour Technol 146:556–561

43. Schievano A, Tenca A, Lonati S, Manzini E, Adani F (2014) Can two-stage instead of one-stage anaerobic digestion really increase energy recovery from biomass? Appl Energy 124:335–342

44. Grimberg SJ, Hilderbrandt D, Kinnunen M, Rogers S (2015) Anaerobic digestion of food waste through the operation of a mesophilic two-phase pilot scale digester–assessment of variable loadings on system performance. Bioresour Technol 178:226–229

45. Zhang J, Loh K-C, Lee J, Wang C-H, Dai Y, Tong YW (2017) Three-stage anaerobic co-digestion of food waste and horse manure. Sci Rep 7:1269

46. Ramakrishnan A, Gupta SK (2008) Effect of COD/NO3−-N ratio on the performance of a hybrid UASB reactor treating phenolic wastewater. Desalination 232:128–138

47. Ferrer I, Serrano E, Ponsa S, Vazquez F, Font X (2009) Enhancement of thermophilic anaerobic sludge digestion by 70oC pre-treatment: energy considerations. J Residuals Sci Technol 6:11–18

48. Choorit W, Wisarnwan P (2007) Effect of temperature on the anaerobic digestion of palm oil mill effluent. Electron J Biotechnol 10:376–385

49. Zhang L, Lee Y-W, Jahng D (2011) Anaerobic co-digestion of food waste and piggery wastewater: focusing on the role of trace elements. Bioresour Technol 102:5048–5059

50. Baêta BEL, Lima DRS, Balena Filho JG, Adarme OFH, Gurgel LVA, de Aquino SF (2016) Evaluation of hydrogen and methane production from sugarcane bagasse hemicellulose hydrolysates by two-stage anaerobic digestion process. Bioresour Technol 218:436–446

51. Liao X, Li H (2015) Biogas production from low-organic-content sludge using a high-solids anaerobic digester with improved agitation. Appl Energy 148:252–259

52. Li Y, Xu F, Li Y, Lu J, Li S, Shah A, Zhang X, Zhang H, Gong X, Li G (2018) Reactor performance and energy analysis of solid state anaerobic co-digestion of dairy manure with corn stover and tomato residues. Waste Manag 73:130–139

53. Chu C-F, Ebie Y, Xu K-Q, Li Y-Y, Inamori Y (2010) Characterization of microbial community in the two-stage process for hydrogen and methane production from food waste. Int J Hydrogen Energy 35:8253–8261

54. Zhu H, Parker W, Conidi D, Basnar R, Seto P (2011) Eliminating methanogenic activity in hydrogen reactor to improve biogas production in a two-stage anaerobic digestion process co-digesting municipal food waste and sewage sludge. Bioresour Technol 102:7086–7092

55. Bolzonella D, Pavan P, Battistoni P, Cecchi F (2005) Mesophilic anaerobic digestion of waste activated sludge: influence of the solid retention time in the wastewater treatment process. Process Biochem 40:1453–1460

56. Wang W, Xie L, Chen J, Luo G, Zhou Q (2011) Biohydrogen and methane production by co-digestion of cassava stillage and excess sludge under thermophilic condition. Bioresour Technol 102:3833–3839

57. Cavinato C, Bolzonella D, Fatone F, Cecchi F, Pavan P (2011) Optimization of two-phase thermophilic anaerobic digestion of biowaste for hydrogen and methane production through reject water recirculation. Bioresour Technol 102:8605–8611

58. Cavinato C, Giuliano A, Bolzonella D, Pavan P, Cecchi F (2012) Bio-hythane production from food waste by dark fermentation coupled with anaerobic digestion process: a long-term pilot scale experience. Int J Hydrogen Energy 37:11549–11555

59. Gianico A, Braguglia CM, Gallipoli A, Mininni G (2015) Innovative two-stage mesophilic/thermophilic anaerobic degradation of sonicated sludge: performances and energy balance. Environ Sci Pollut Res 22:7248–7256

60. Mejías R, Rubén E (2013) Optimization of biogas production in two-stage anaerobic fermentation of organic waste market using alkaline pretreatment, Tesis at Politecnico di Torino. Turin, Italy

61. Liu D, Min B, Angelidaki I (2008) Biohydrogen production from household solid waste (HSW) at extreme-thermophilic temperature (70°C)—influence of pH and acetate concentration. Int J Hydrogen Energy 33:6985–6992

62. Badroldin NA (2010) Treatment of palm oil mill effluent (POME) using hybrid up flow anaerobic sludge blanket (HUASB) reactor

63. Tauseef SM, Abbasi T, Abbasi SA (2013) Energy recovery from wastewaters with high-rate anaerobic digesters. Renew Sustain Energy Rev 19:704–741

64. Zhang L, Ban Q, Li J, Jha AK (2016) Response of syntrophic propionate degradation to pH decrease and microbial community shifts in an UASB reactor. J Microbiol Biotechnol 26:1409–1419

65. Li Y, Gao M, Hua D, Zhang J, Zhao Y, Mu H, Xu H, Liang X, Jin F, Zhang X (2015) One-stage and two-stage anaerobic digestion of lipid-extracted algae. Ann Microbiol 65:1465–1471

66. Moset V, Ottosen LDM, Xavier C de AN, Møller HB (2016) Anaerobic digestion of sulfate-acidified cattle slurry: one-stage vs. two-stage. J Environ Manage 173:127–133

67. Puebla YG, Pérez SR, Hernández JJ, Renedo VS-G (2013) Performance of a UASB reactor treating coffee wet wastewater. Rev Ciencias Técnicas Agropecu 22:35–41

Printed in the United States
by Baker & Taylor Publisher Services